GLOBAL SURVIVAL

"CHANGE THE WORLD"

Where there is no vision, the people perish.
—Psalms, Old Testament

*One cannot solve a problem with the same
kind of thinking that gave rise to the problem.*
—Albert Einstein

This series of books written by and for the Club of Budapest is addressed to all people; women and men, young and old, wherever they live and whatever their interests and lot in life. It is published in the conviction that today's world must be changed, and that it can be changed—and that how we think and what we do are key to changing it.

It is evident that we need new thinking. Our world has become both socially and ecologically unsustainable, and continuing on the present path would lead to growing crises and, inevitably, to catastrophe. This is our "problem"—and this problem, as Einstein said, cannot be overcome with the same kind of thinking that led to it. Without a vision of where we want to go, we will continue to reach for old and outdated objectives; and, if we persist, ultimately we will surely perish.

We also need new ways of acting. We cannot pursue our own life and interests as if they were separate and disconnected from the lives and interests of others. We now live in a global family—the planet is our home and our village. We need planetary vision in thinking and planetary ethics in acting.

Planetary vision and planetary ethics are not utopian abstractions. They are the most concrete challenges we have ever faced, whether as individuals or as a species. They are the preconditions of our wellbeing and our development, and even of our survival.

In practical terms and in everyday language, the books published in the CHANGE THE WORLD series discuss new ways of thinking and acting, and how all of us can achieve them in our daily life.

GLOBAL SURVIVAL

The Challenge and
Its Implications for
Thinking and Acting

Ervin Laszlo and Peter Seidel, Editors

Foreword by Ervin Laszlo

SelectBooks, Inc.

Global Survival: The Challenge and Its Implications for Thinking and Acting
© 2006 by Ervin Laszlo and Peter Seidel

This edition published by SelectBooks, Inc., New York, New York.

First Edition

ISBN 1-59079-104-5

Library of Congress Cataloging-in-Publication Data

Global survival : the challenge and its implications for thinking and acting / Peter Seidel and Ervin Laszlo, editors.
       p. cm.
   Includes bibliographical references and index.
   ISBN 1-59079-104-5 (hardbound : alk. paper)
   1. Sustainable development. 2. Globalization. I. Laszlo, Ervin, 1932–
   II. Seidel, Peter, 1926–
   HC79.E5G597145 2006
   338.9'27–dc22
                                2005030813

Manufactured in the United States of America

10 9 8 7 6 5 4 3 2 1

Contents

Foreword

I welcome Peter Seidel's initiative in bringing together a group of outstanding thinkers to address the topic of Survival Research. This is indeed a most timely topic, the groundwork for a promising and much needed new discipline.

The world today is unsustainable—it threatens not only the well-being of the privileged minority, but the survival of the whole species. The degeneration of the local and global environment, the inequities of the globalized socioeconomic system, and the stress, misery, and resentment which now exist among billions of people have created a situation that is either quickly and effectively remedied or leads to a series of breakdowns. The September 11 terrorist attacks on New York and Washington were not the cause of crisis in the world: the crisis is due to the way the world's socioeconomic system is structured and managed.

The current socioeconomic system of the world has concentrated production, trade, finance, and communication in the hands of a few, and has created misery coupled with national and regional unemployment, widening income gaps, and an impoverished environment for many. While the richest 20% of the world population becomes richer still, the poorest 20% is pressed into abject poverty, whether in rural areas or in shantytowns and urban ghettos. These conditions fuel resentment and revolt, and provoke massive migration—from the countryside to the cities, and from the poorer to the richer regions. Fanatics wage holy wars and resort to terrorism, and organized crime engages in information fraud, corruption, and traffic in women and children as well as drugs, organs, and weapons. In industrialized countries personal as well as job security are things of the past. In poor countries poverty is aggravated by hunger, joblessness, and degrading conditions. Both rich and poor countries overwork productive lands, contaminate rivers, lakes, and seas, and draw down water tables. And the gap between the modern and the traditional segments of society rends apart the structures and institutions on which social stability depend.

Albert Einstein noted that we cannot solve a problem with the same kind of thinking that gave rise to that problem. Today our problems are global, and the new thinking we need must likewise spread over the globe. What humanity urgently needs to overcome the threat to individual and collective survival is not heavy-handed intervention from the top down, but new thinking percolating from the grass roots up. This, in turn, requires a better flow of relevant information based on a sober assessment of what is wrong with the world today, and what each of us can do to help put it right.

For this reason we must give urgent consideration to the proposal made in this volume: to create an open-ended dialog centered on what we need to, and can, do to ensure the survival of our species on this precious but vulnerable and already dangerously overloaded home planet.

Ervin Laszlo
June 14, 2005

Introduction

A Sensible Way to Think and Act

Today we are threatened in many ways—by war with weapons of mass destruction, environmental degradation, and a wide variety of social, political, and economic problems. We know the dangers, yet we allow them to worsen. Continuing on this path can only result in vast misery and, possibly, even the extinction of our species. Simply put, we are winning some of the battles but losing the war. We can and must do better to correct these problems, not just lessen their symptoms. We need to continue to learn more about them, discover the best ways to deal with them, learn why we so often fail to take adequate steps to protect our future, and then find the most effective ways to resolve them.

A new way for viewing the world and dealing with it would help. While we know much about many of the problems we face, such as rampant population growth, soil depletion, the contamination of aquifers, and our growing vulnerability to various kinds of social breakdown and terrorism, we know too little about how these interact with each other and the possible cumulative consequences of those interactions. We must not only examine all of the known dangers and their possible interactions, but must look for those we are not yet aware of and encourage investigation of them.

We need to focus on relationships between the human mind, society, and the Earth's biosphere. We must learn how our brains, technology, governments, businesses, beliefs, ethical systems, and our evolutionary development interact. Instead of looking ever more inward, we must look outward—searching for connections, gaps, and unasked questions. By examining the big picture we can not only devise ways to deal with physical problems, but also practical ways we can bridge the limits of our minds, society, organizations, and governments, to circumvent the obstacles they present and overcome properties inherent in them that are causing us harm or preventing us from taking effective action.

We must reach out both horizontally and vertically. Horizontally we must look for connections between a broad variety of disciplines and ways of thinking. We must connect what we already know, and ask for study in areas where we know little. When considering the future prospects of world food production, we must consider not only population growth, but the future outlook for energy, agricultural chemicals, water, soils, climate change, ecology, political systems and their stability, disease, transportation, business, ethics, psychology, and a host of other subtly related issues. Vertically, we must examine problems and mechanisms from the abstract right down to where things get done and, where needed, develop ways to do better.

Some years ago I read John Herz's unpublished paper proposing Survival Research.* Survival Research would do all of those things. I was impressed, but occupied by other things at the time. I was disturbed that a presentation he had made in 1988 was never followed up on, and that the paper I had read had not been published. I was also surprised that what seemed so logical and so much needed had not blossomed out in a formal manner elsewhere. (Since the special issue of *World Futures* containing these articles was published, I discovered that it has. See: http://sustsci.harvard.edu/index.html.) Thinking that something should be done, I made copies of Herz's paper and sent them with a letter to a number of individuals I thought should be interested. I received a reply from Ervin Laszlo, the founder of the discipline of systems philosophy and former president of the International Society for Systems Sciences. He responded by inviting me to guest edit an issue of *World Futures,* of which he is editor. I could not resist. It grew to be a double issue, and now in addition, the book you are reading.

The chapters in this book suggest the breadth that must be covered, though they touch only some of the areas needing investigation. There are no articles on ecology, business, communications, or public education, for example. What they do is demonstrate the range of factors that influence the sustainability and quality of life. Many multinational corporations, whose only legal responsibility is to their stockholders, are larger, richer, and more powerful than many national governments. These businesses, with staffs of public relations experts and access to open global mar-

---

* Survival as used here covers the range from survival in a partly damaged environment to the actual extinction of our species.

kets, interact with the environment, governments, societies, and the lives of each one of us—often to our detriment. We know a lot of things that have nothing to do with survival, and are ignorant of many things that are essential for it. This affects how we treat our environment.

The authors of the papers presented here have addressed their subjects as they see them relating to the sustainability and quality of life. Some writers may present similar thoughts from different viewpoints, and some ideas may conflict. That is unimportant. Our purpose here is to generate interest in and encourage the development of what Herz calls Survival Research, and to demonstrate the need to reach out in all directions. The mere existence of this field, whether it is called Survival Research or Sustainability Science, will draw attention to the extent of the problems we face and the need for investigation along these lines.

The papers that follow present some thoughts of my own, a rewritten, updated version of John Herz's original paper on Survival Research based on his work developing the idea in the 1980s, and the thinking of scholars from a variety of disciplines. David and Marcia Pimentel examine the interactions between population, natural resources, and life-support systems. Walter Lowen explains how the structure of the brain affects how we deal with the world around us. Jerome Barkow reveals how our evolutionary past governs the way we think today. James Alcock describes our need to believe, and how our beliefs influence our behavior—and our planet. David Myers shows us that when one has the basic physical and psychological necessities of life, more wealth and consumption do little to increase happiness. Ervin Laszlo discusses the need for a planetary ethic and proposes criteria for establishing one. Joseph Tainter discusses a framework for sustainability and suggests a way to achieve it. Richard B. Norgaard and Paul Baer recommend ways to integrate knowledge that is now divided between poorly communicating disciplines, much to our detriment. Kenneth Watt notes powerful forces that combine to restrict the spread of a systematic approach to survival and suggests solutions. Andy Bahn and John Gowdy demonstrate how neoclassical economics harms Earth's life-support system, and call for a more inclusive view of reality. J.R. McNeill describes how a better understanding of history can provide the perspective we need to recognize our current predicament. Richard Lamm asks whether democracy can resolve the new set of survival problems we face, and presents ways to do better. Christopher Williams discusses how leaders can acquire the necessary attributes to address the threats that face us. Lindsey Grant examines foresight in government. James Lovelock proposes that we produce an easily accessible civilization survival manual for future generations.

The material in this volume originally appeared as volume 59, numbers 3–4, April–June 2003 issue of *World Futures*. The authors were here given the opportunity to update and revise where they felt it was necessary. Recent events show that the message presented then is needed even more today. We are indebted to Muriel Adcock, managing editor of the U.S. editorial office of *World Futures*, for her tireless work at bringing this out as a double issue of *World Futures*, and later for her help in getting it published as a book.

Peter Seidel

1

Survival Research: A New Discipline Needed Now

Peter Seidel

Today we face a challenge that no species has had to face before. We have to look far into the future and, as best we can, assume responsibility for it. In the past the laws of evolution and ecology maintained order—at times brutally. For us this has changed; through our cleverness we have broken the constraints on our ecological niche, allowing our numbers and their impact on the Earth to expand at the expense of other forms of life. We have fouled our surroundings, produced weapons that can destroy most life on Earth, and formed civilizations that, unlike animal or pre-agriculture cultures, enable a few to live in extravagant luxury while many do not even have shelter or enough to eat. We are on our own as a species, responsible for our behavior. However, we have not accepted this responsibility, preferring instead to fight among ourselves and allow our primitive drives to dominate our actions.

When we focus on recent and possible future developments—as in medicine, for example—we can see a bright future. However, when we broaden our view to include the increasing cost of medical care and drugs, the expanding number of elderly and impoverished people worldwide, the shortage of doctors in rural and undeveloped areas, the growing resistance of many bacteria to antibiotics, the rapid spread of diseases due to air travel, tropical diseases appearing in northern climates as a result of global warming, and more floods and hurricanes spreading disease, the situation is impossible to predict. Add to this inevitable wars, economic downturns, and social chaos.

Likewise, consider how the following would affect optimistic predictions for feeding the world's people: the topsoil in food exporting nations such as the United States continues to deteriorate at current levels, farmland is lost to other uses, pests continue to become immune to pesticides, fossil fuels (the raw material for important chemical fertilizers) become scarce and expensive, water supplies for irrigation diminish, and the world's population increases significantly. If the United States cannot produce enough food for its own people, how will nations dependent on importing food cope—particularly those without cash? Will still higher-yield varieties of crops, more vulnerable to disease and thirsty for water, dependent on pesticides, herbicides, chemical fertilizers, and energy, be able to overcome these problems?

There are many other factors that will affect our future, for example: rampant poverty, an increasing disparity between the rich and the poor, pandemics, charismatic tyrants with easy answers, hate, racism, nationalism, religious and political fanatics, terrorism, prolific and deadly micro weapons, hungry talents willing to sell their minds to the highest bidder, corruption, moral breakdown, and despair.

As we head into the future, new discoveries and inventions will appear that will help us in significant ways we do not yet know. But trusting science, technology, and unrestricted free enterprise and free trade to solve our problems is unwise. They cannot create more living space and cropland for a population that will continue to grow, replenish depleted aquifers, nor resurrect extinct species with their learned behaviors.

Need for a New Discipline

What we need is an eyes-wide-open holistic view of where we are, what we are doing, where we are headed, why, and how we can do better. A discipline devoted to this end (potentially named "Survival Research") would provide us with information that would enable us to act responsibly as a species.

This idea is not new. It was introduced by the noted political scientist John Herz at a conference sponsored by the Ralph Bunche

Institute on the United Nations at the Graduate Center Of The City University of New York in March 1988, at which scholars from various disciplines discussed threats to the future existence of humankind. Herz's proposal, in the form of an article, was submitted to a political science journal but was rejected in spite of the author's reputation. Although the idea did not receive adequate attention, it was important then and is even more so now.

Herz suggested that "survival research must rise above the specific concerns, interests, even expertise, of any particular discipline, such as political science. It must be interdisciplinary, requiring the cooperation of any and all the social sciences with other scientific disciplines, such as agronomists, climatologists, physicists, medical sciences, and so forth ... It must mobilize experts in the various fields so as to make them recognize the superdisciplinary concerns of global survival to which priority must be given over and above (and, possibly, in contrast to) the more parochial concerns ...."[1]

Although not proposing a new discipline, in his book *Consilience* entomologist Edward O. Wilson calls for synthesis and cooperation. "Most of the issues that vex humanity daily—ethnic conflict, arms escalation, overpopulation, abortion, environment, endemic poverty, to cite several most persistently before us—cannot be solved without integrating knowledge from the natural sciences with that of the social sciences and humanities. Only fluency across the boundaries will provide a clear view of the world as it really is ... To the extent that the gaps between the great branches of learning can be narrowed, diversity and depth of knowledge will increase ... The enterprise is important for yet another reason: it gives ultimate purpose to intellect. It promises that order, not chaos, lies beyond the horizon."[2] Wilson suggests, "the world henceforth will be run by synthesizers, people able to put together the right information at the right time, think critically about it, and make important choices wisely."[3]

Survival Research

Quoting from my own book, *Invisible Walls*, "survival studies would be an extension of ecology, concentrating on interactions between the human mind, society, and the Earth's biosphere. We must understand the effect of our lifestyle on the world around us and learn how our minds, government, business, religion, and our evolutionary development interact. Excellent work has already been done in this direction in such subjects as general systems theory, general living systems theory, evolutionary psychology, sociobiology, and ecological economics. This trailblazing needs to be connected, coordinated, and expanded."[4]

Survival Research would evaluate what society most needs to know now, and encourage study in these areas. It would also examine how we determine what data is to be sought out; how information is stored, retrieved, connected, and integrated; and how it moves us into action. It would ask crucial obvious, and not so obvious, questions that simply are not asked (or if so, answered only superficially). For example, assuming that the Earth's population cannot continue to grow perpetually, by whom, when, and how will it be stopped, and whose responsibility is it to do so? And, if active steps are not taken, what are the possible consequences?

This new discipline would examine how existing political and economic systems interact with each other. It would run long-term scenarios on how world trade could affect the environment, societies, cultures, communities, and individuals. It would ask, what happens if a country that holds a large portion of a world resource, say petroleum, decides that it is in their interest to develop it slowly? How might the petroleum-hooked nations of the world respond, and how can the world's people be better prepared for such an event?

In *Invisible Walls* I examine why we do so poorly at resolving problems of which we are aware and cognizant, and are able to deal with safely. This enigma resides in our minds, society, organizations, governments, beliefs, and ethical systems. Investigating obstacles to reasonable action and ways to overcome them would be an important task for Survival Research.

Survival Research Would Propose Solutions

Besides providing information and telling us where we are, Survival Research would suggest ways for humans to live more harmoniously with each other and our planet. It would develop ways we can bridge the limits of our minds, societies, organizations, and governments, and avoid the destructive quirks inherent in them.

Think of it. Most of the problems we have today are the result of humanity's past efforts to better itself. The automobile has provided us with a wonderful means for moving about, but it is damaging the Earth's atmosphere and many of the communities we move about in and leaves many people dead or maimed. While many saw television as a boon for education, in a number of ways it has turned out to be the opposite. Today, many of our intentions to do good things work against each other or create new problems. This is the consequence of single-minded thinking, which can achieve narrowly focused goals most easily. Resolving problems in ways that contribute to the resolution of other problems, and devising means to further cooperation within and between organizations and governments, would direct significant amounts of human energy away from counterproductive activities and toward truly productive ones. And, by utilizing humanity's greatest talents and intelligence in government and elsewhere, instead of ignoring them as we often do, we would amplify these gains.

Survival Research would look for ways to insert constructive feedback loops into organizations, including governments. Raising taxes on fossil fuels would reduce their use and encourage conservation. Taxing stock trades would encourage people to look for sound long-term investments, reduce speculation, and raise money from parties that benefit from government regulation. Survival Research would look for ways to restructure governments to make them operate more efficiently and democratically. It would produce exemplary models of governments which would meet such requirements. Finding ways to give citizens a clearer picture of issues and personalities, and to get communally involved individuals who understand about human and planetary needs to run for office, would strengthen democracy and further good governance.

A reasonably accurate world model (acceptable to most nations and religions) produced by respected international scholars would help establish a common ground between the world's people, allowing us to reach shared goals. Society would also gain by making economic models more realistic—by recognizing that many resources are finite, for example. A group of respected leaders, theologians, and philosophers from different countries, religions, and cultures might work together to produce a set of ethical principles that most people could endorse. Widespread publicity given to these agreed-upon principles would affect public opinion, put pressure on political leaders, and reduce dangerous conflict.

Call for Knowledge

Once, all creatures knew what they needed to know to survive and procreate. Most knew little more. Today we know a lot of things that have nothing to do with survival, and are ignorant of many things that are essential for it. Besides providing a more realistic view of reality, exploring problems in our dealings with reality, and suggesting ways to do better, Survival Research would establish goals for what teachers, journalists, leaders, and every one of us should know in order to be good citizens of our communities, nations, and planet. This knowledge would include a basic understanding of the workings of our individual minds, society, government, and the physical world.

Because the subject it deals with is so important for future generations, Survival Research would not be a value-free discipline. Where it sees a need, it would advocate action. One of its activities would be to solicit money from governments, foundations, businesses, and individuals to fund a broad array of educational programs for teachers, journalists, and public officials.

Over our history, human concerns have gradually branched out from ourselves and our children to our extended family, tribe, and nation-state. But we now must take a bigger step. Although it may not come naturally to us, we must broaden our concerns to include all humanity and life; not just for now, but on into the distant future as well.

Too often we know what needs to be done, but do not do it. This is true for individuals, and even more so for society. We are concerned about the future of our species, but are doing a very poor job of addressing this head on because we deal with problems as unrelated fragments. We fight battles instead of wars, and alleviate symptoms rather than deal with causes. It would help greatly here to have an acknowledged discipline, open-ended and directed toward integrating all aspects of survival inquiry that individuals and institutions could address themselves to. It would provide a focal point where ideas could be exchanged and propagated, finances sought, and projects funded. Survival Research has the potential to move us beyond our current impasse of winning some battles while we lose the war.

Notes

1 Herz, John. Some observations on engaging in survival research. 2002. *International Journal of Applied Economics and Econometrics.* Jan.–March.
2 Wilson, Edward O. 1998. *Consilience,* New York: Alfred A. Knopf. p. 13.
3 *Ibid.* p. 296.
4 Seidel, Peter. 1998. *Invisible Walls.* Amherst, New York: Prometheus Books.

2

On Human Survival

Reflections on Survival Research and Survival Policies

John H. Herz

This contribution to *Global Survival* is a revised and, in the post-scripts, updated rendering of an idea first proposed in the mid-1980s and later presented as an address by the author at a conference at the Graduate Center of the City University of New York in March 1988, held under the auspices of its Ralph Bunche Institute on the United Nations. Participants from various disciplines discussed serious threats to the very future of mankind, such as the threat of extinction through all-out nuclear war or through the confluence of over-population, the overuse and destruction of vital resources, and environmental deterioration. In that address the author contended that it is not only necessary that experts in their own diverse disciplines and areas of investigation recognize these problems, but that they work on them in interdisciplinary cooperation with each other. The author believes such interdisciplinary research and consequent international action have now become even more urgent even where, because of the end of the Cold War, the threat of nuclear extinction appears to have diminished. Note, for example, the threats of nuclear proliferation in south Asia (India and Pakistan), the Middle East, and Korea, as well as climate change (global warming).

Survival Research as proposed here deals with the question of how human survival in a reasonably livable environment can be ensured in the face of threats, ranging to possible extinction, which confront the entire human race. It is not concerned with the "survival" of specific groups within mankind, such as formerly that of the West vs. the East or Israel vs. the Arabs, but rather with the novel situation where, for the first time in human history, the future of every one of us is in jeopardy. The global survival threat arises from two primary areas, both resulting from the tremendous upsurge of new scientific discoveries and technological developments: 1) that from nuclear weapons and the ensuing danger of an all-out nuclear war; and 2) that of the combined effects of the ongoing population explosion, the destruction of the human habitat, pollution, and depletion of vital resources. (In the following, I shall call the latter the environmental or ecological threat, because it affects the very inhabitability and carrying capacity of our planet.)

Considering our predicament, philosopher Hans Jonas suggested the following "categorical imperative" for actors in international relations, if not for all of us: "Act in such a fashion that the effects of your action are not destructive of the future possibility of life on earth."[1] In the mid 1980s at the annual CUNY Political Science Conference, I made the following statement urging professional involvement with the survival issue: "I would like to take the occasion of our Political Science Conference to propose to the profession the establishment of a new subdiscipline, on the model of Area Studies, or the recently emerging Holocaust Studies: Survival Research. If a holocaust that cost the lives of six million deserves a special scholarly approach (and it certainly does deserve it), the threat of a universal holocaust should cause on the part of professionals—political scientists and others—special effort to study ways and means to prevent it. Survival Research should be given a prime place on our scholarly agenda."

George Kennan, a foremost scholar-statesman of the twentieth century, concurs by drawing attention to two major universal dangers and stressing the urgency for dealing with them: "... our world is at present faced with two unprecedented and supreme dangers. One is the danger not just of nuclear war, but of any major war at all among great industrial powers—an exercise which modern technology has made suicidal all around. The other is the devastating effect of modern industrialization and overpopulation on the world's natural environment ... Both are relatively new problems, for the solution of which past experience offers little guidance. Both are urgent. The environmental and nuclear crises will brook no delay. The need to give priority to the averting of these two over-riding dangers has a purely rational basis—a basis in national inter-est—quite aside from morality."[2]

The nuclear threat, involving one sudden and overwhelming occurrence, was the first to become obvious to us. The ecological one takes longer to sink in, if it does at all, as it is the composite of a large number of gradually, although often exponentially, emerg-ing developments whose interrelatedness is not immediately clear. (For example, life-threatening climatic change resulting from industrial and vehicular carbon dioxide or similar emissions.) Nevertheless, it would seem to require only a small effort to see such trends in their connectedness and thus detect the survival threat that the sum-total of these ecological trends poses. It should also be easy to understand, as Kennan advocates, that both threats should be given priority over all other matters of policy, because extinction or drastic harm to the human race renders all issues of "victories" or "prevalences" nugatory. Safeguarding our future is the overriding issue. However, scanning the daily news or television screen reveals how little interest there is in survival issues. The question is, why?

There are a number of reasons for the neglect. I emphasize five of them:

First, the novelty (or, by now, non-novelty) of the threats, notably that of the nuclear bomb. Impressive as it was, and still is, for those of us who are old enough to remember Hiroshima and Nagasaki, it is less so for those born into the atomic age. They have learned how to "live with the bomb" and consequently may feel less urgency to "do something about it." Likewise, we have all become used to living with the ecological threat, and even get bored with constantly hearing about toxic pollution of this or that. Even dying lakes or woods are nothing novel for those who have grown up with these events.

Second, older individuals retain the habit of concentration on tra-ditional parochial concerns within one's own group's narrow inter-ests, inducing the nation and its "national interests" to be narrowly conceived. Petrified ways of thinking, based on equally fixed ways of perceiving, are not easily replaced by thinking in terms of global sur-vival and perceiving the world as one. The simplistic idea of a defen-sive "shield" protecting a nation against a nuclear attack, based on antiquated concepts and realities of defensibility, may prevail even in the face of the new phenomenon of nuclear penetrability of even the most powerful nations. If national leaders and prominent scientists like Edward Teller apparently are unable to free themselves from antiquated perceptions, how can we expect the man in the street to know better? After the first atomic explosion, it is said that Albert Einstein remarked, "The unleashed power of the atom has changed everything save our modes of thinking; thus we drift toward unparalleled catastrophe."

Third, as terrifying as the nuclear threat might be in our imagination, for most of us it remains invisible; except for the few people who live near the installations, we just do not see it with our eyes as being real. The Pentagon used to present newly elected American Presidents with nuclear realities by showing them movies of nuclear explosions and their consequences, thus impressing upon them the gravity of their responsibilities regarding "pushing the button." I am not aware whether this is still done. Since Reagan became president, policies have not reflected a sense of awareness.

Regarding the ecological threat, the slowly evolving nature of change makes these dangers less visible to us. Every child born into the world contributes to the population explosion; however, its cumulative effect is not noticeable at any one moment. The same applies to the overuse of water supplies and energy resources, overfishing, deforestation, the pollution of "just one more little stream or lake," and so on. While people may know about such phenomena and their effects, they rarely grasp their impact on the global ecology.

Fourth, there is the growing impact on public opinion by economically and/or ideologically focused groups interested in the maintenance of a favorable status quo that, while endangering the future of the world, promotes their narrows interests. Businesses interested in a deregulated economy free from government interference and environmental regulations, and the military interested in job security guaranteed by continually increasing armament programs, are examples of this.

In our era of "public relations," vested interests utilize ever more sophisticated means of influencing, if not controlling, public opinion and attitudes. This partly explains the failure of movements and organizations concerned about threats to global survival to impart this awareness to larger publics or to mobilize for more than short periods of time. Policy is occupied more with creating comforting public images than with accomplishing results such as resolving problems through mutual compromise. We must then ask, are catastrophic events such as a nuclear attack or another Chernobyl needed to bring us to our senses? How much radioactive material do we have to absorb to comprehend the threat, and to understand that it transcends national and regional boundaries? Tragically, an even greater prod may be needed to provoke action regarding threats to global survival. Instead, as an article in the *New York Times* once put it, "presidents have failed to educate the nation, preferring instead to apply a cheerful veneer ... Leaders fear being tagged as scare-mongers or doomsayers. Optimists are cast as winners, pessimists as losers."

Fifth, it may not be special or parochial interests purposely deflecting the public from recognizing survival-threatening factors, such as using low-mileage SUVs. It may simply be the unintended non-awareness of such connections. Even environmentalists and other reform-minded groups may, when advocating reforms, mention only the more immediate consequences of what they advocate, such as more comfort, better health protection, and so forth, instead of pointing out the more profound but less immediately apparent connections to survival. Thus, gas-guzzling motor vehicles not only pollute the air and potentially contribute to increased incidences of lung cancers (as does the unlimited emission of carbon dioxide from power plants), but they contribute to global warming. But somehow mentioning such more remote but far-reaching effects in their connectedness with daily lifestyles or ordinary production methods has become taboo. It is not politically correct. But exactly here lies the responsibility of survival researchers to clarify the relationship between causes and effects, "ordinary" and customary attitudes and actions and destructive survival consequences.

Survival Research, in order to arrive at conclusions that are practicable in policies, not only must recognize the above listed impediments and suggest how best to overcome them, but should also suggest attitudes and approaches that, while differing from inherited traditional patterns, are appropriate to the solution of new and unfamiliar survival problems. In the following, I mention a few that come to mind:

1. While in ordinary, non-survival situations "worst case" scenarios can be ruled out as too unlikely, they cannot be disregarded when global survival is at stake. In the new age, as Hans Jonas puts it, "the prophecy of doom is to be given greater heed than

the prophecy of bliss," and "not even for saving his nation's life may the statesman use means that can destroy mankind."[3]

The stakes are now too high. Thus, in the field of nuclear deterrence one cannot, or should not, rely on hair trigger computerized warning systems. Starting from "worst case" assumptions, Survival Research has to probe under which circumstances the effects can be avoided (complete nuclear disarmament, low maximum level of nuclear armament, etc.).

2. In contrast to the nuclear threat, the great complexity inherent in the various facets of the ecological threat makes it less easy to arrive at Survival Research-type conclusions. As I have pointed out, specific developments such as deforestation and desertification in sub-Saharan Africa or overpopulation in Third World metropolitan areas may not appear as dangers to the global ecosystem, per se. Only in their totality will they turn out to be life-threatening for all of us. Thus, the industrialization of Third World countries in conjunction with the absence of effective population controls has meant resource depletion, urbanization without raising the living standards of the majority of the population, and migration of rural populations to megacities. This causes the destabilization of regimes that try to defend themselves against overthrow through import of arms (at the expense of resources for genuine development) and incessant, endemic internal or external, minor or major warfare, often accompanied by the laying of landmines that make entire regions uninhabitable. The overall world arms production and trade, and the ensuing armaments and arms races, not only raise the specter of nuclear war through escalation but, with their tremendously accelerating waste of non-renewable resources, pose a direct threat to ecological survival. This affects not only Third World countries, but the industrialized countries as well. The ever-growing portion of GNP deflected to military purposes results in a vast waste of human and material resources, neglect of infrastructure, erosion of the welfare state and ensuing economic and social polarization, and so on.

3. Survival Research has to analyze these trends in all their complexity, and then make suggestions for safeguarding the human future in the face of all the threats involved. Such research must confront the basic underlying problem of the compatibility (or incompatibility) of existing economic, social, and cultural/reli-

gious systems, and the interests of the elite in such systems, with the "survival" aim of a balanced, or rebalanced, ecology of the globe. To put it the other way around: what are the economic, social, etc. requirements of a livable world? Survival Research must rise above the specific concerns, interests, even expertise, of any particular discipline, such as political science. It must be *interdisciplinary,* requiring the cooperation of any and all the social sciences with other scientific disciplines, such as psychologists, demographers, agronomists, biologists, climatologists, physicists, medical sciences, and so forth, even theologians. In doing so, it must mobilize experts in the various fields so as to make them recognize the *superdisciplinary* concerns of global survival to which priority must be given over and above (and, possibly, in contrast to) the more parochial concerns of this or that national, economic, religious, or similar grouping. But where to find them, whom to address?

In one of my writings I referred to the international civil service as a potential "universal class" in the Hegelian sense of a group outside and thus above the various societal "interests," a class able to rise above considerations of mere parochial concern.[4] Since then we have, of course, witnessed a tremendous increase in agencies and organizations directly or less directly affiliated with the United Nations, with a corresponding increase in personnel employed by these agencies and in the U.N. itself. Whether many of them have turned out to be true "universalists" is doubtful. Even the international civil servant tends to be "parochial" not only in his or her affiliation with the country of origin, but also regarding his or her professional specialization. Yet it is clear that there is a vast reservoir of expertise in global matters such as food supply, population trends, energy resources, and so forth that can and should be made use of by survival researchers.

The same applies to scientists of all disciplines within national environments. In our age of global survival concerns, it should be the primary responsibility of scholars to engage in survival issues. We no longer can afford the luxury of "value-free" research in the Weberian sense. As I, quoting myself in my statement to the political science conference mentioned above, put it:

"Because in today's world, for the first time, the survival of all is in jeopardy; even those who, like myself, are value relativists (i.e., believe that, in principle, no 'Ought' can be derived from an 'is') can agree that, when certain values become so over-whelmingly important that their non-recognition appears absurd to practically everybody engaging in human discourse, those values can be posited as certain, or undeniable. Where the alternative to the 'Ought' denotes physical extinction of the entire human race, survival not of individuals or specific groups, but of mankind as such, becomes an absolute value."[5]

Here too, and even to a larger extent than that of international organizations, there is the likelihood and danger that too many prove too much beholden either to a specific regime or to special (e.g., military or business) interests.

As for material to be used by survival researchers, there is, of course, an abundance of same, collected by scientific organizations. This is in addition to the wisdom accumulated at innumerable conferences of international agencies on demographic and other ecological issues, not to mention the enormous material compiled by the United Nations itself.

The point is to have all this accumulated material as well as the conclusions drawn from it by Survival Research trickle down to a concerned and attentive public. Awareness of issues and priorities is crucial. Thus, a special responsibility of survival scholars lies in the education of those who have a special impact on public world-awareness: teachers in the primary and secondary schools and in the colleges, writers and journalists in the press and electronic media. Above all, survival information and awareness must be made available to the incoming generations of elementary, high school, and college students. Hence the prime importance of university teachers conveying a sense of urgency of the survival problems to all those who otherwise might, like lemmings, follow the call to the precipice.

Postscript 1

Looking back upon the last century that accompanied my long life, I see it as a century of missed opportunities. The first was the failure of the League of Nations to replace a system of ever more destructive wars with one of peace through collective security. After its failure, it appeared that the new United Nations system, with its agreement on the non-use of force, might initiate a new era of peace. But after the emergence of the nuclear bomb, bipolar competition of two ideologically opposed superpowers rendered the U.N. collective security system inapplicable.

The greatest failure of opportunity, however, came with the end of the Cold War of the two superpowers. With the disappearance of the Soviet regime it seemed that no enemy threatened the democratic Western world any more, and thus collective peace policies based on the U.N. system could prevail in international relations. As far as armaments, including nuclear ones, were concerned, one could expect a "peace dividend" in the form of drastic reduction of armaments, including nuclear ones, and vastly diminished expenses that would make means available to deal with the other survival problems.

Instead, everything has been going in the other direction. Instead of reduction of armaments and "defense" budgets, we have tremendously rising armaments and costs; instead of nuclear disarmament agreements, we see the spread of the weapons to countries like India and Pakistan, engendering nuclear crises that threaten global survival. Incredibly, under the one remaining superpower's new nuclear policy review, one considers not only "preemptive" use of nuclear weapons against another nuclear power, but even such use against non-nuclear countries, and envisages novel "small" nuclear bombs to be used against underground installations.

And with an act of terrorism by nonstate actors interpreted as an "act of war," anti-terrorist action threatens to render the normal international state of peace a state of permanent war, with the threatened superpower claiming the right unilaterally—that is, independently—to act wherever it claims that "rogue" states or "rogue" regimes threaten security. Quite generally, being part of international bodies, international treaties, even commitment to binding rules of international law or the jurisdiction of international courts, is considered as not in one's national interest.

But what may be called a global obsession with terrorism has not only affected the survival area of mass destructive weapons, but has deflected attention from the other areas of global survival—the ecological ones. One may almost call it a case of collective amnesia. Who, especially among leaders of nations, still worries about the annual one hundred million human beings added to the global population and their effect on Third World living standards and global resources? Who is still concerned with global warming and related climate change?

There has been a world-wide change of ideology and policies from a search for domestic and international regulation and planning of "better worlds" (that since the age of industrialization created welfare states and measures of environmental protection and preservation) to the ideology and policies of the regulation-free market system, with non-interference of state and international organization in the affairs of globalized economies. Laissez-faire means non-pursuance or even abrogation of efforts and measures toward environmental protection. In the United States, the Bush regime has been abrogating whatever modest measures had been taken by preceding administrations, whether in regard to diminishing the emission of carbon dioxide gases or the destruction of forests or other land in the interest of free exploitation by timber or oil-drilling interests, not to mention reducing the already insufficient aid to Third World nations to create even minimally sufficient infrastructures for sustainable human habitat. Without a Marshall Plan-like program of vastly increased aid, terrorism will continue to be used as political expression, but any such effort must be combined with efforts to deal with overpopulation, i.e., to reduce family size, without which aid efforts would be "down the rat hole."

Survival Research must also be aware of priorities. There is a present-day obsession with health matters, to the neglect of other survival problems. Of course, it is of great importance to pursue health policies such as combating epidemics (such as AIDS) all over, especially in the Third World. But what use is it, for instance, to devote limited human and material resources to reducing infant mortality where surviving children have nothing to look forward to but joblessness, misery, and a slum-life in poverty?

Generally speaking, the laissez-faire ideology in an age of rampant, unyielding globalization implies a tendency on the part of vested interests to spread this ideology and the corresponding attitudes to the general public through control of the media and, increasingly, even the educational institutions forming the minds of the new generations. This even extends to influencing academia.

While I am writing these lines, two non-official gatherings of world leaders in business, politics, and culture have concluded their sessions: the Davos World Economic Forum, held in New York, and a parallel World Social Forum at Porto Allegro, Brazil. I am not aware that even the latter, held in opposition to Davos, was interested in survival problems like global warming or the population crisis and the ensuing need of family planning. Shouldn't we be interested in an annual "Global Survival Forum" assembling survival researchers?

In view of the foregoing, is there still hope? I remember that I had little hope when, decades ago, I ended a book dealing with survival matters with the words that classical playwrights used to put at the end of their play manuscripts: *"Exeunt omnes—Finis."* But pessimism does not necessarily mean fatalism. In contrast to facile optimism based on illusory utopian expectations, pessimism may be more realistic. To give an example from my own life: if, as a German Jew, I had embraced the optimism that engaged many fellow Jews in those early years after Hitler's rise to power, I would have ended in a gas chamber. My more realistic pessimism made me emigrate in time and survive. And thus, at the end-stage of my life, I refuse to give up hope for human survival on this beautiful, challenging planet.

This hope, however, is based on one all-important requirement: the turn around of attitudes and policies from individual and group concerns to global awareness of what has to be done for continued survival of the species *Homo sapiens,* a turn around that demands leadership to persuade millions to change lifestyles and make the sacrifices needed for survival. Radical turns have happened in human history before—now it must happen globally. If the species is not able collectively to act so as to provide for its continued existence, then indeed its final fate will be indicated by the words *Exeunt omnes—Finis.*

Postscript 2

The publication of this book gives me a chance, for the second time, to write a postscript to the original version of my presentation on human survival problems. Only a few years after the publication of my first postscript, a preemptive war perpetrated by the self-styled supreme world power, and that war's effect on the system and relations of powers, has made it clear to me that understanding what is happening requires a deeper understanding of human history.

All species survive through the instinct of self-preservation. But *Homo sapiens* has been unique among all other species by being able to act consciously, by applying thought, reasoning, and evaluation to his problems. Thus *Homo sapiens* has created organized societies, civilizations, nation states, and in this has tremendously improved his living standards. It has not meant, however, that with all the glorious achievements of the thousands of years long history of the *sapientes* there has not been oppression, discrimination, cruelty, constant war, even extermination. True, wise and compassionate men time and again have searched for and propagated the "good life," the physically and morally better ways of behavior and action; thus the ancient Jewish prophets and the great philosophers of ancient Greece. But, until relatively recently, they have been without much effect on actual human life. Then something novel happened. Beginning three to four centuries ago, Enlightenment ideas and movements have managed to create real progress in systems and societies. Terrible effects of early industrialization were improved by progressive movements such as labor movements, and then peace movements, influencing nation states to establish international organizations and live up to at least minimally effective rules of international law. The welfare states overcame the worst features of unmitigated capitalism; emancipation did away with slavery; Jews and other traditional targets of discrimination were admitted to equal status; decolonization meant at least the beginning of equality between the former colonizers and the vast populations of the Third World. True, there were repeated attempts to return to the old pre-Enlightenment ways—Hitlerism, Stalinism. But the general trend was that of liberalism, progressivism, and humanitarianism. And this trend could well become the basis for an effective approach to the solution of survival threats.

Do the attitudes and policies of the Bush regime merely oppose this or that specific feature of the Enlightenment era, or are they rather a fundamental alteration in modern history? I am afraid it is the latter. It is an ideologically compact whole of "true believers," believers in a social Darwinism where the economically powerful have the right to exploit the "unfit," and the politically powerful the weak, on the primitive basis of a Manichaean distinction between "good" and "evil," the good and the bad guys, friends and enemies.

The situation perhaps can be best discussed under the heading of two laissez-faires, the economic and the political. The first considers a completely unregulated free trade and free market system as the only effective system, thus putting labor under the control of corporate power. Everything is directed toward economic "growth," which, of course, is needed, especially for rapidly increasing populations, but does not mean much as long as it provides prosperity for wealthy elites at the expense of the purchasing power of the mass of people. These remain impoverished. Above all, growth in the Third World—one has said ironically—is also the ideology of the cancer cell. One can observe those cells in the slums surrounding the megacities of the Third World.

When I took a course in economics as a student in pre-Nazi Germany, each economics professor, whether capitalist or socialist, defined economics (Volkswirtschaft) as the study of determining which was the best way to satisfy the basic needs of a country's inhabitants; that is, not only how to have the GNP grow, but also how to distribute it among all the people. The distribution question now is neglected by the theorists (for example, the Chicago school economists) and by government professionals.

The ideological agenda of the Bush regime thus aims at the complete reversal of the Enlightenment agenda, going back to years when property was not interfered with by limitations and regulations of the welfare state or, more recently, by rules for environmental protection. With globalization of markets and economic competition, it has already meant exporting jobs or entire industries to countries of low wages and living standards (outsourcing), increasing unemployment and reducing workers' income in the United States itself, and increasing the gap between the super-rich and the poor, with ever more billionaires enjoying and consuming needed resources, confronting the ever rising number of the poor all over the world. And with global competition, the property ideology compels other developed countries still pursuing welfare policies to follow suit. And this renders nugatory pursuance of environmental agendas, and therewith attempts to solve survival issues likewise, whether one tries to prevent property interests from pollution of air, water, or land, or forests from timber or oil-drilling interests.

The second laissez-faire is that of international and military action, where the Bush regime's refutation of even basic rules of international law surpasses the era of Hugo Grotius *(pacta sunt servanda)*. The invasion of Iraq was in violation of the United Nations charter rules permitting use of force only in self-defense or by permission of the Security Council. Its treatment of prisoners of war ("hostile detainees") disregards the Geneva Convention and the treaty forbidding torture. And the jurisdiction of international courts is opposed or disregarded.

In the nuclear arms field, American unilateralism interferes with the aim of preventing nuclear catastrophes. Of course, further nuclear proliferation should be inhibited. But what about arms reduction by the original nuclear powers? The Nuclear Non-Proliferation Treaty was concluded by promising those not-yetnuclears that the nuclear possessors would disarm too. The United States is now doing the opposite. Bush has even cancelled the bilateral agreement with Russia forbidding missile defense, an agreement that had made it easier to rely on mutual deterrence from a nuclear first strike. Now, free to build a missile defense system, he is likely to immensely increase security fears of all but the United States, and is thus likely to compel those who feel threatened to become nuclear or engage in increased nuclear arms races.

In yet another withdrawal from arms control, President Bush—having withdrawn from a decades-old treaty that declares outer space as outside any country's sovereignty—is now attempting to establish American "supremacy" in outer space by producing and placing there what one calls "space weapons." What next? U.S. sovereignty over the moon? Mars? The rest of the solar system?

But it appears that the wars undertaken by the United States "against terrorism" are not even successful in their objectives. Instead of "regime change" to increase "democracy," they seem to simply create more chances for terrorists, bringing chaos instead of liberal democracy. It appears that war, instead of maintaining or extending super-power "supremacy," can no longer be successful when the enemy no longer fights with organized armed forces, but with terrorists ready to commit suicide.

The lack of success, so far, of a war based on the social Darwinist ideological claim to have the right to wage war wherever and whenever it serves the interest of the super-power to preserve its military and political supremacy has not prevented the Bush regime from undermining the prestige of the United Nations, and therewith the chances of peaceful solutions of conflicts. The world organization and the members backing its force-restraining rules have officially declared the unilateral use of force illegal, but instead of punishing the perpetrator they are being "punished" by the superpower for not joining it in its use of force.

The worldwide public reaction against the war in Iraq, including anti-war sentiment in the United States, has induced the regime to declare peace and freedom to be its major foreign policy agenda. Indeed, "law and order" have actually been aims of the regime's policy, but not so much in the interest of the respective countries as in that of investors (American and multinational) in making sure that their economic interests are protected in a stable political climate, with laws and courts protecting property. Propagation of peace is needed to calm public opinion aroused by the war and the cruelties and sufferings it has brought about. This is true for politics and opinion at home, too. Where one considers oneself in a permanent state of "war against terrorism," ever more concentration of power in the executive renders opposition ineffective and endangers a democracy based on the balance of political power among the legislature, judiciary, and executive. It also endangers chances of enacting environmental protections, and therewith survival-ensuring measures. But assuring regime-favorable public opinion is likewise helped by an alliance with religious fundamentalism—the Christian Right. While its vales are not generally social Darwinist, there is the common anti-abortion interest and a host of related issues. But with increased Muslim immigration, in the U.S. and in Europe, are we thus increasing religious confrontations, possibly even religious war?

What are the chances of global survival movements under these circumstances? I said global survival *movements*. In the face of the historical turn away from the progressive enlightenment attitudes and policies of the last centuries, an equally strong and worldwide turn back to them must be the basis for measures effectively dealing with the great survival issues. Just to mention a few: stopping global warming through drastically reducing carbon dioxide emissions requires drastic changes in lifestyles and economies (no SUVs, cleaner and small and fewer cars, or bicycling instead, greatly increased gas taxes, as in Europe); nuclear power plants replacing coal and oil-fed ones; prohibiting fishing in large areas of the high seas; drastic reduction of family sizes in Third World countries, which will require having social security systems because families will have fewer children only when they can be sure that they will be cared for in old age.

The latter example shows that the rich countries will have to give tremendous aid, which they can afford by cutting down their vast arms budgets. (This also would help in solving another survival problem, that of nuclear disarmament and arms control.) This requires international, even global, action and cooperation through international organization and institutions such as the United Nations; and great sacrifices—of economic advantages, of national "sovereignty," of individual satisfactions and enjoyments.

Is it realistic to expect such radical changes in attitudes and policies world-wide, of readiness by people and nations to make the necessary sacrifices? But radical turn-arounds have happened before.

Think of the United States' own turn from laissez-faire and isolationism to the New Deal and the establishment of a world-wide peace organization in the years 1933–45. But it must be a true "movement"—compelling parties, politicians, and leaders to engage in popular aims. And environmentalists must make sure that such a movement, besides being neo-liberal and a peace movement, must also be an environmental one. We must do all in our power to see that the survival issues are made clear to citizens, voters, readers of newspapers, media watchers, and especially the incoming generation (courses in survival issues in high school and colleges, etc.). Realizing how strongly the "neo-cons" or the Murdochs have managed to control much of public opinion, it all depends on intellectual leaders, liberals, environmentalists, to gain such influence. But even mentioning the problems has often been taboo, even on the part of liberals, and especially as liberal parties have become more centrist, hardly remembering their progressive past (Labor in England, the Democrats in the United States.) We must remind them or change them. (Imagine "regime change" in the greatest polluter of all, the United States!) There have been beginnings of living up to the Kyoto requirements in Europe, for instance; or Brazil taking measures to protect what is left of the greatest rain forest in the world. Recently, some cities in the United States, frustrated by the Bush policy, have enacted ordinances in an effort to implement some of the provisions of the Kyoto treaty. But how long can countries be expected to make sacrifices while the richest and largest remains inactive or even spoils the process?

Have I engaged in utopianism? I have always been fighting utopian visions, but considered realizable agendas to be in conformity with political realism. A realistic liberalism can promote progressive and humanitarian ideals as well as survival aims. On this, my long-held theoretical basis, I still, at the end of my long life, consider survival agendas as realizable. But each year our rising number makes this beautiful planet more uninhabitable—not only for us, but for many other species. Be active. It's up to you.

Acknowledgments

I wish to thank Peter Seidel, urban planner and author of the most comprehensive recent book on global survival problems (*Invisible Walls*, 1998), for having encouraged me to offer a now over fifteen year-old talk for publication. It may be outdated in parts, but I believe that the issues it deals with are as urgent as they were in 1988, if not more so. On this, see my article "Reflections on my Century."[6] I also wish to thank my son Stephen Herz and, in particular, my colleague and friend Professor Tom Karis, without whose assistance this version of my talk could not have been accomplished.

Notes

1 Jonas, Hans. *The Imperative of Responsibility: In Search of an Ethics for the Technological Age.* Chicago: University of Chicago Press, 1984: p. 11.
2 Kennan, George.1985/86. Morality and foreign policy. *Foreign Affairs.* Winter: p. 216.
3 Jonas, Hans *op. cit.* p. 36ff.
4 Herz, John H., *International Politics in the Atomic Age.* New York: Columbia University Press, 1959: p. 320. I expected a "universalist attitude" to emerge, through a kind of "self-election" in "those who are gripped by a sense of responsibility and concern for the future and who, at the same time, possess the broadness of vision which is required to grasp the common survival interest of mankind." Ibid. p. 327.
5 Herz, John H. 1976 Technology, ethics, and international relations. *Social Research.* Spring: p. 107.
6 Herz, John H. 1998. Published as Occasional Paper in 1998 by the Ralph Bunche Institute on the United Nations at the Graduate Center of the City University of New York; And, 2002. *International Journal of Applied Economics and Econometrics.* Jan.–March: pp. 151–163.

3

World Population, Food, Natural Resources, and Survival

David Pimentel and Marcia Pimentel

World and U.S. Population Growth

The current world population stands at more than 6.1 billion. It has doubled during the last 45 years, and based on the present growth rate of 1.3% per year, the population is projected to double again within a mere 50 years (Figure 1).

Many countries and world regions have populations that are rapidly growing. For example, China's present population is 1.3 billion and, despite the governmental policy of permitting only one-child per couple, it is still growing at an annual rate of 1.2%.[1] But China, recognizing its serious overpopulation problem, has recently passed legislation that strengthens its one-child per couple policy. However, because of the young age structure of the Chinese population, it will continue to increase for another 50 years. India, with nearly 1 billion people living on approximately one-third the land of either of the United States or China, has a current population growth rate of 1.9%, which translates to a doubling time of 37 years.[2] Taken together, the populations of China and India constitute more than one-third of the total world population. Given the steady decline in per capita resources, it is unlikely that India, China, or the total world population will double.

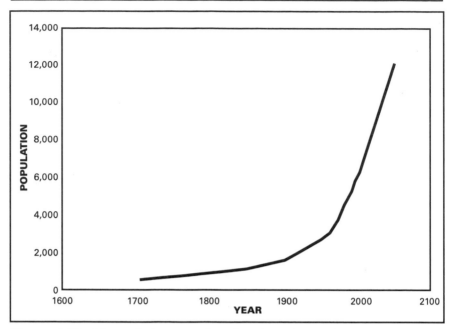

Figure 1. World population growth and projected growth based on the current population growth rate of 1.3% per year.

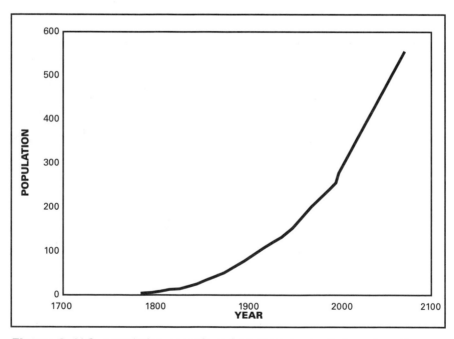

Figure 2. U.S. population growth and projected growth based on the current population growth of about 1% per year.

Also, the populations of most African countries are also expanding. For example, the populations of Chad and Ethiopia have high rates of increase and are projected to double in 21 and 23 years, respectively. Similarly in Latin America, the populations of Paraguay and Mexico have high population growth rates that are projected to double in 26 and 37 years, respectively. In contrast, the European populations are increasing, but at extremely slow rates.[3]

The United States population also is growing rapidly and currently stands at 285 million, having doubled during the past 60 years (Figure 2). Based on its current growth rate of about 1% it is projected to double to 570 million in just 70 years (Figure 2).

A major obstacle in limiting human population growth is the very young age structure of the current populations and the population momentum that it fosters. With ages that range from 15 to 40, reproductive rates are high.[4] Even if all the people in the world adopted a policy so that only an average of two children were born per couple, it would take approximately 70 years before the world population would finally stabilize at approximately 12 billion, which is twice the current level[5, 6] (Figure 3). As the world and U.S. populations continue to expand, all vital natural resources will have to

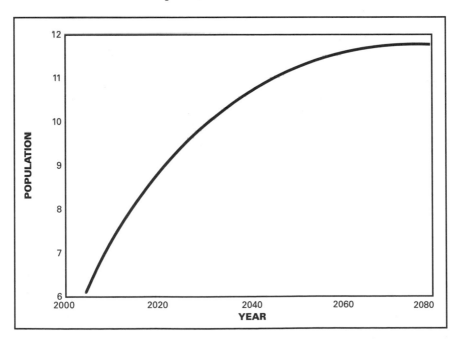

Figure 3. Population project stabilization assuming a 2-child average per couple adopted in 2000 (Weeks, 1986; Population Action International, 1993).

be divided among increasing numbers of people and per capita availability will decline to unacceptably low levels. When this occurs, maintaining prosperity, quality life, and personal freedoms will be imperiled.[7]

Malnourishment in the World

The present world hunger and shortages of nutrients for many humans alerts us to an existing serious problem concerning the world food supply and its impact on human health. The report of the Food and Agricultural Organization (FAO) of the United Nations confirms that food per capita has been declining since 1984, based on available cereal grains (Figure 3). This is alarming news because cereal grains make up about 80% of the world's food supply. Although grain yields per hectare in both developed and developing countries are still increasing, the rate of growth is slowing while the world population and its food needs escalate.[8,9] Specifically from 1960 to 1980, U.S. grain yields increased at about 3% per year, but since 1980 the annual rate of increase for corn and other major grains has declined to only about 1% (Figure 4).

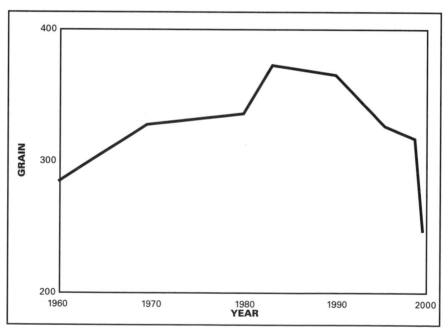

Figure 4. Rate of increase for U.S. grain yield, 1960–2000.

According to the World Health Organization, more than 3 billion people are malnourished.[10,11] This is the largest number and proportion of malnourished people ever reported! The World Health Organization, in assessing malnutrition, includes deficiencies of calories, protein, iron, iodine, and vitamin A, B, C, and D shortages in its evaluation.[12,13]

World Cropland Resources

More than 99.7% of human food comes from the terrestrial environment while less than 0.3% comes from the oceans and other aquatic ecosystems.[14,15,16] Worldwide, of the total of 13 billion hectares of land area on Earth, the percentages in use are: cropland, 11%; pasture land, 27%; forest land, 32%; urban, 9%; and other, 21%. Most of the remaining land area (21%) is unsuitable for crops, pasture, and/or forests because the soil is too infertile or shallow to support plant growth, or the climate and region are too cold, dry, steep, stony, or wet.[17,18]

Per Capita Cropland

In 1960, when the world population numbered only 3 billion, approximately 0.5 ha of cropland per capita was available. This area

Resources	U.S.	China	World
Land			
Cropland (ha)	0.50	0.08	0.25
Pasture (ha)	0.83	0.33	0.55
Forest (ha)	0.92	0.11	0.73
Total (ha)	**3.25**	**0.52**	**2.13**
Water (liters x 10^6)	1.7	0.46	0.64
Fossil fuel			
Oil equivalents (liters)	8,000	700	1,800
Forest products (kg)	1,091	40	70

[a] USDA (2001); [b] USBC (2000); [c] PRC (1994); Bennett (1995); [d] SSBPRC (1990); [e] Buringh (1989); [f] International Energy Annual (2001); [g] UNEP (1985).

Table 1. Resources used and/or available per capita per year in the U.S., China, and the world to supply basic needs.

is considered essential to produce a diverse, healthy, nutritious diet of plant and animal products—similar to the typical diet of people living in the United States and Europe.[19,20]

Even now, as the human population continues to increase and expand its diverse activities, including transport systems and urbanization, vital cropland is being covered and lost from production. Globally, the annual loss of land to urbanization and highways ranges from 10 to 35 million hectares (approximately 1%) per year, with half of this lost land area coming from cropland.[21] As a result, the average per capita cropland worldwide has diminished to about 0.25 ha, or about half the amount needed to provide diverse food supplies similar to those enjoyed in the U.S. and Europe (Table 1). In the U.S. the average cropland per capita is now down to 0.5 ha, or the critical area essential for food production.[22]

China is an example of the rapid changes occurring in the availability of per capita cropland. There the amount of available cropland is only 0.08 ha per capita. This relatively small amount of cropland provides the Chinese people primarily a vegetarian diet (Table 2). Chinese cropland is rapidly declining due to continued population growth, but also because of extreme soil erosion and degradation.[23]

Food/feed	**U.S.**	**China**	**World**
Food grain	92	387	158
Vegetables	191	198	167
Fruits	135	35	60
Meat and fish	91	62	56
Dairy products	272	7	79
Eggs	15	14	8
Fats and oils	31	5	13
Sugar and sweeteners	72	7	25
Nuts	9	NA	1
Total food	**908**	**406**	**567**
Feed grains	816	70	150
Kcal/person/day	3,800	2,734	2,808

[a] USDA, 2001; [b] FAOSTAT, 2001; [c] Wan Baorui, 1996.

Table 2. Foods and feed grains supplied per capita (kg) per year in the U.S., China, and the world.

The availability of cropland influences the kinds and amounts of foods produced. For example: currently, a total of 1,481 kg/yr per capita of agricultural products is produced to feed Americans, while the Chinese food supply averages 785 kg/yr per capita (Table 1). By all available measurements, the Chinese have reached or exceeded the limits of their agricultural system.[24] Furthermore, their reliance on large inputs of fossil-fuel based fertilizers to compensate for shortages of arable land and severely eroded soils, plus their limited fresh water supply, suggests severe problems looming in the near future.[25] Even now China imports large amounts of grain from the United States and other nations, and is expected to increase imports of grains in the near future.[26]

In summary, the growing shortage of productive cropland is one of the underlying causes of the current worldwide food shortages and poverty in many regions of the world.

Loss of Cropland

In addition to the intrusion of humans and their activities on the Earth's land area, degradation of soil has emerged as a critical agricultural problem.[27,28] Throughout the world, current erosion rates are greater than ever previously recorded.[29,30]

Worldwide, more than 10 million hectares of productive arable land are severely degraded and abandoned each year.[31,32] To compensate, about 10 million hectares of new land must be put into production each year to attempt to feed the human population. Most of the additional cropland needed yearly to replace lost land comes from the world's forest areas.[33,34] This urgent need accounts for more than 60% of the massive deforestation now occurring worldwide.[35] Estimates are that approximately 1 ha each of forest and pasture land per person is a reasonable goal.

Soil erosion degrades and diminishes the productivity of cropland and pasture land.[36,37] The primary causes of this degradation are rainfall and wind erosion, as well as the salinization and waterlogging of irrigated soils.[38]

The current high erosion rate throughout the world is causing loss of productivity. When soil is eroded, valuable soil nutrients, soil water resources, and soil organic matter are lost, and soil depth is reduced as well. Topsoil renewal is extremely slow. In fact, it takes approximately 500 years for 2.5 cm (1 inch) of topsoil to form under agricultural conditions.[39,40,41,42]

Soil erosion on cropland ranges from about 13 tons per hectare per year (t/ha/yr) in the United States to 40 t/ha/yr in China.[43,44] During the past 30 years, the rate of soil loss throughout Africa has increased twenty-fold.[45] Worldwide, soil erosion averages approximately 30 t/ha/yr, or about 30 times faster than its replacement rate.

Under some arid conditions in India, and with relatively strong winds, as much as 5,600 t/ha/yr of soil has been reported lost.[47] Related to wind erosion, during the summer of 2001 the National Aeronautical Association photographed an enormous cloud of soil being blown from the African continent toward the South and North American continents. Wind erosion is so serious in China that its soil particles are detected in the Hawaiian atmosphere during the annual Chinese spring planting.[48]

In addition, slope and intensity of rainfall intensify soil erosion. For instance, in the Philippines, on sloping land under tropical rainfall, as much as 400 t/ha/yr of soil can be lost.[49] Furthermore, these large amounts of eroded soil pollute streams and rivers, as evidenced by the reported 2 billion tons/year of soil transported down the Yellow River of China.[50]

Soil scientists, including Myers, estimate that approximately 75 billion/tons per year of soil are lost from world agricultural lands.[51] Then, too, in many developing countries soil erosion is intensifying because fuel wood is in short supply and people burn crop residues as a fuel. Removal of crop residues leaves soils unprotected from the effects of wind and rainfall, thereby intensifying erosion.

Agriculturists know that the fertility of nutrient-poor soil can be improved by large inputs of fossil fuel-based fertilizers. This practice, however, increases dependency on the finite fossil fuels stores used in the production of fertilizers. Even with current fertilizer use, soil erosion remains a critical problem in current agricultural production.[52]

Crops can be grown under artificial conditions using hydroponic techniques, but the costs in terms of energy expenditure and dollars is approximately ten times that of conventional agriculture. Such systems are not sustainable for the future.

Water Resources

All life requires significant amounts of fresh water. The total amount of water made available by the world hydrologic cycle is sufficient to provide the current world population with adequate fresh water. Yet world water supplies are concentrated in some areas, while other areas are short of water or outright arid.

Water for Human Consumption

The minimum basic water requirement for human health, including drinking water, is considered to be about 50 liters per capita per day.[53] The U.S. average for all domestic usage, however, is eight times higher than that figure, or 400 liters per capita per day.[54] Americans use water freely in their agriculture, homes, and gardens.

Rapid human population growth and associated increased water demand already is stressing the world's water resources. Worldwide, between 1960 and 1997, the per capita availability of freshwater worldwide declined by about 60%. Another 50% decrease in per capita water supply is projected by the year 2025.[55] Water demands already far exceed supplies in nearly 80 world nations.[56] For instance, in China more than 300 cities suffer from inadequate water supplies, and the problem is quickly intensifying as the population increases.[57,58] In arid regions such as the Middle East and parts of North Africa, where yearly rainfall is low and irrigation is expensive, the future of agricultural production is grim.

Water for Food Production

All vegetation requires enormous quantities of water during the crop growing season. For example, an average corn crop that produces about 8,000 kg/ha of grain uses more than 5 million liters/ha during its growing season. To supply this much water to the crop, approximately 1,000 mm of rainfall per hectare must reach the plants. If irrigation is required, about 14 million liters of irrigation water is required during the growing season.[59]

Sources of Water

Surface Water: Rainfall provides all the water found in streams, rivers, lakes, and oceans, and it is a vital part of the hydrologic cycle. Frequently surface water is not managed effectively, resulting in water shortages and pollution, both of which threaten humans and the aquatic biota that depend on it. The Colorado River, for example, is used so heavily by Colorado, California, Arizona, and other adjoining states that by the time it reaches Mexico it is usually no more than a trickle running into the Sea of Cortes.[60] Also, water shortages and pollution are experienced in the lower Colorado

River and are a major threat, having destroyed many important aquatic species that inhabit that part of the river.

Groundwater: Rainfall is also stored in enormous underground aquifers. Their slow recharge rate from rainfall is usually only between 0.1% and 0.3% per year.[61,62] At such a slow recharge rate, groundwater resources must be carefully managed to prevent overuse and depletion, but this is not the case. For example, in Tamil Nadu, India, groundwater levels declined 25 to 30 meters during the 1970s because the pumping of irrigation water was excessive. [63,64] Similarly in Beijing, China, the groundwater level is falling at a rate of about 1 m/yr; while in Tianjin, China, it drops 4.4 m/yr.[65]

In the United States, ground water overdraft is high, averaging 25% greater than replacement rates.[66] An extreme case is the vast Ogallala aquifer located under Kansas, Nebraska, and Texas. There the annual depletion rate is 130% to 160% above replacement.[67] If these rates continue, this large aquifer, which supports agricultural irrigation and countless communities in central U.S., is expected to become non-productive by 2030.[68]

Rapid population growth and increased total water consumption combine to rapidly deplete water resources. The present and future availability of adequate supplies of freshwater for human and agricultural needs is already critical in many regions, especially the Middle East and parts of the North Africa where low rainfall is endemic.[69]

Irrigation

Irrigation enables crop production to succeed in arid regions, provided there is an adequate source of fresh water. Currently, approximately 70% of the water removed from all sources worldwide is used solely for irrigation.[70,71] Of this amount, about two-thirds is consumed by growing plant life and is non-recoverable.[72] An irrigated corn crop requires about 14 million liters/ha of water and uses about three times more energy to produce the same yield as rain fed corn.[73,74] Also, the costs of irrigation are extremely high, with the irrigation per hectare costing about $1,200.

The limitation of surface and ground water resources for irrigation, and its high economic costs plus heavy energy inputs, will tend to limit future agricultural irrigation. This will be especially true in developing nations where economics cannot support such expenditures.

Many times irrigation results in salinized and waterlogged soils, both of which diminish crop productivity. These problems result from continuous irrigation, especially when there is either insufficient water and/or poor soil drainage hindering the flushing of the salts from the soil.[75] In addition, subsurface runoff and leaching from saline soil increases salt levels in river water. This happens as the Colorado River flows through the Grand River Valley in Colorado and water is withdrawn for irrigation. Some irrigation water is later returned to the river, along with an estimated 18 t/ha of salts leached from the soil.[76]

Unfortunately, it is not possible to use the vast oceans of water when fresh water sources are in short supply. Desalinization is an energy intensive process and is economically impractical on a large scale. For example, the amount of desalinized water required by one hectare of corn—14 million liters—would cost about $14,000, while all other inputs, like fertilizers, cost only $600.[77, 78] This water cost does not include the additional expense of moving vast amounts of water from the oceans to the distant agricultural fields.

Water Pollution

A major threat to maintaining ample fresh water resources is pollution. Although considerable water pollution has been documented in the United States,[79] this problem is of greatest concern in countries where water regulations are not rigorously enforced or do not exist. This is common in developing countries that discharge approximately 95% of their untreated urban sewage directly into surface waters. For instance, of India's 3,119 towns and cities, only 209 have partial sewage treatment facilities and a mere eight possess full waste-water treatment facilities.[80] A total of 114 Indian cities dump untreated sewage and cremated bodies directly into the sacred Ganges River.[81] Downstream, the polluted water is used for drinking, bathing, and washing.

As mentioned, serious water pollution problems exist in the United States as well. EPA reports indicate that 37% of U.S. lakes are unfit for swimming because of runoff pollutants and contamination from septic discharge.[82] In addition, pesticides, fertilizers, and soil sediments pollute water resources when they accompany eroded soil into water bodies.

Also, industries all over the world often dump untreated toxic chemicals into rivers and lakes. Thus, pollution from sewage and disease organisms, as well as the 100,000 different chemicals used globally, makes water unsuitable not only for human drinking but also for application to crops.[83]

Eroded soil sediments washed into reservoirs during rainfall creates another problem. Estimates are that about 1% of the volume of reservoirs is being filled with sediments each year, thereby reducing the volume of water available for irrigation and other purposes.[84] The total cost of sedimentation plus the loss of the water storage capacity of each dam worldwide is estimated to be about $7.5 million per year.[85] It would be economically impractical to dig and remove the soil sediments deposited in the reservoirs.

Energy Resources

Humans have relied on various sources of power for centuries, those energy sources having ranged from human, animal, wind, tidal, and water energy, to wood, coal, gas, oil, and nuclear sources for fuel and power. Since about 1700, abundant fossil fuel energy supplies have made it possible to augment agricultural production to feed an increasing number of humans. In addition, human travel, trade, and the transportation of food and manufactured goods that enhance human life have benefited from fossil energy. Energy availability has made it possible to purify and transport water and by plant production has helped the production of thousands of different drugs and pharmaceuticals.[86] All these energy-based improvements have enhanced human quality of life.

Energy Source	U.S.	World
Petroleum	37.7	138
Natural gas	22.1	86
Coal	21.7	100
Nuclear power	7.7	34
Biomass	7.0	29
Hydroelectric power	3.4	27
Geothermal and wind power	0.4	0.8
Biofuels (ethanol)	3.4	7.0
Total consumption	**96.6**	**412.8**

[a] USBC, 2000 (thermal equivalents for hydropower and nuclear power); [b] International Energy Annual, 2001; [c] Estimated, Pimentel, unpublished.

Table 3. Fossil and solar energy use in the U.S. and world (quads).

In essence, ample energy supplies—especially fossil energy—have supported rapid population growth, industrialization, transportation, urbanization, plus high consumption rates (Table 3). Worldwide about 412 quads (1 quad = 10^{15} BTU, 0.25 x 10^{15} kcal, or 433 x 10^{18} Joules) from all energy sources are used each year.[87] In fact, the rate of energy use from all sources has been growing even faster than world population growth. Thus, from 1970 to 1995, energy use increased at a rate of 2.5% per year (doubling every 30 years), compared with the worldwide population growth of 1.7% per year (doubling about 40 years).[88,89] During the next 20 years, energy use is projected to increase at a rate of 4.5% per year (doubling every 16 years) compared with a population growth rate of 1.3% per year (doubling every 54 years).[90,91]

Although about 50% of all the solar energy captured by worldwide photosynthesis is used by humans, it is still inadequate to meet all of human needs for food and other needs.[92] To make up for this shortfall, about 348 quads of fossil energy, mainly oil, gas, and coal, are utilized worldwide each year for all activities.[93] Of this, 83 quads are utilized in the United States.[94]

Each year, the U.S. population uses 50% more fossil energy than all the solar energy captured by harvested U.S. crops, forest products, and other vegetation.[95] Industry, transportation, home heating, and food production account for most of the fossil energy consumed in the United States.[96,97] Per capita use of fossil energy in the United States is 8,000 liters of oil equivalents per year, more than twelve times the per capita use in China (Table 1). In China most fossil energy is used by industry, although a substantial amount, approximately 25%, now is used for agriculture and the food production system.[98,99]

Taken together, developed nations annually consume about 70% of the fossil energy worldwide, while the developing nations, which have about 75% of the world population, use only 30%.[100] The United States, with only 4% of the world's population, consumes about 24% of the world's fossil energy output (Table 3).

Some developing nations that have especially high rates of population growth are increasing fossil fuel use to augment their agricultural production of food and fiber. For example, in China since 1955 there has been a 100-fold increase in fossil energy use in agriculture for fertilizers, pesticides, and irrigation.[101]

In general, fertilizer production has declined by more than 17% since 1989, especially in the developing countries, because of fossil fuel shortages and its high prices.[102] In addition, the overall projections of the availability of fossil energy resources for fertilizers and all other purposes are discouraging, as the stores of these finite fossil fuels decline.

Fossil Fuel Supplies

The world supply of oil is projected to last approximately 50 years at current production rates.[103,104,105,106] The worldwide natural gas supply is considered adequate for about 50 years, and coal for about 100 years.[107,108,109] These estimates, however, are based on current consumption rates and current population numbers. If all people in the world could enjoy a standard of living and an energy consumption rate similar to those of the average American, and the world population continued to grow at a rate of 1.5%, the world's fossil fuel reserves are estimated to last only about 15 years.[110,111]

Petroleum geologist W. Youngquist reported in 1997 that current oil and gas exploration drilling data had not borne out some of the earlier optimistic estimates of the amount of these resources yet to be found in the United States.[112] Both the production rate and proven reserves have continued to decline. Domestic oil and natural gas production have been declining for more than 30 years and are projected to continue to decline.[113] Approximately 90% of U.S. oil resources already have been mined.[114] At present, the United States is importing about 61% of its oil. This dependency puts the U.S. economy at risk due to fluctuating oil prices and difficult political situations, such as the 1973 oil crisis and the 1991 Gulf War.

With the exhaustion of fossil fuels and associated increases in costs and pressure from global climate change, significant changes will also have to take place in energy use and practices. Fossil fuel shortages and global warming problems will force a transition to renewable energy sources in the future.

Renewable Energy

Although many solar technologies have been investigated, most are only in limited use. The most promising of renewable sources of energy include: solar thermal receivers, photovoltaics, solar ponds, wind-power, hydropower, and biomass energy.[115]

By using available renewable energy technologies such as biomass energy and wind power, an estimated 200 quads of renewable energy per year could be produced worldwide. However, to do this would require from 20% to 26% of the land area for solar energy production.[116] A self-sustaining renewable energy system producing 200 quads of energy per year for about 2 billion people would provide each person with 5,000 liters of oil equivalents per year. Although this is about half the average American consumption rate per year, it would be an increase for most people of the world. However, the requirement to use over 20% of the land area (excluding cropland) for renewable energy production could limit the resilience of the vital ecosystem that humanity depends upon for its food and other life support systems.[117]

Biodiversity

In addition to energy resources, crop, forest, and livestock species, human life depends on the presence and functioning of approximately 10 million plant, animal, and microbe species existing in agroecosystems and throughout the natural environment.[118,119] Although approximately 60% of the world's food supply comes from rice, wheat, and corn species, as many as 20,000 other plant species are used by humans to some extent for human food.[120,121]

The natural ecosystem and the diverse species it contains serve as a vital reservoir of genetic material for future development in agriculture, forestry, pharmaceutical products, and biosphere services. Yet with each passing day an estimated 150 species are eliminated from the planet because, as human numbers continue to increase, their diverse activities expand into natural ecosystems. These activities include deforestation, soil and water pollution, pesticide use, urbanization, development and extension of transport systems, and industrialization.[122] The rate of extinction of some species under these conditions is 1,000- to 10,000-times faster than occurs in natural systems.[123] The fact that humans use more than 50% of the solar energy captured by the entire plant biomass to meet their food and fiber needs is a major cause of these high rates of extinction. This means that the photosynthetic biomass available to maintain vital natural biota is significantly reduced and biodiversity is greatly reduced.

Humans have no technologies which can substitute for the food, medicines, and diverse services that plant, animal, and microbe species provide. For example, one third of the human food supply relies either directly or indirectly on effective insect pollination.[124] Each year, honey bees and wild bees are essential in pollinating about $40 billion worth of U.S. crops, in addition to natural plant species. Including pollination, the economic benefits of biodiversity in the United States are an estimated $300 billion per year, and nearly $3 trillion worldwide. Indeed, plants, animals, and microbes also carry out many other essential activities for humans, recycling manure and other organic wastes, degrading some chemical pollutants, as well as purifying water and soil.[125]

Many living organisms assist in crop protection. Approximately 99% of crop pests are controlled by diverse natural enemy species, and the development of pest resistance in host-plants is assisted by genetic material that comes from wild plants found in natural ecosystems.[126] Research efforts need to be focused on using diverse natural enemies and the genetics of host-plant resistance to enhance pest control. The development and use of biological controls along with sound ecological techniques like crop rotation will help reduce pesticide use and, ultimately, costs to farmers.[127]

Most agriculture worldwide depends on introduced species of plants and animals. However, some introduced species are causing extinctions of native species worldwide.[128,129] In addition, the invasive species are having negative impacts on the food production and economies of most nations. For instance, in the U.S. invasive species are causing more than $137 billion in damages each year.[130]

Despite the yearly use of 3 million tons of pesticides and other controls worldwide, pest insects, pathogens, and weeds destroy about 40% of all potential crop production.[131] Specifically in the United States, approximately 0.5 million tons of pesticides are applied each year, yet pests still destroy about 37% of all potential crop production.[132]

Environmental pressure imposed by the expanding the human population is the prime destructive force in reducing biodiversity. Yet programs for world biological conservation are focused on protecting species in national parks that cover only 3.2% of the world's terrestrial area.[133] Certainly this conservation is helpful, but overlooked is the fact that humans currently occupy 95% of the terrestrial environment, either as managed agricultural and forest ecosystems or human settlements.[134] This means that most species diversity occurs in managed terrestrial environments, and emphasizes the need for greatly enhanced efforts to improve the sustainability of world agricultural and forest ecosystems.[135]

Natural Resources and Human Diseases

Human health may seem unrelated to natural resources, but upon closer consideration it becomes apparent that both the quality and quantity of all natural resources (e.g., food, water, and biodiversity) play a central role in maintaining human health. As populations increase in size, risks to health and productivity grow as well, especially in areas where sanitation is inadequate. In view of rapid population growth, it is not surprising that human deaths due to infectious diseases increased more than 60% from 1982 to 1992.[136,137,138]

These increases in diseases are associated with diminishing quality of water, air, and soil resources, and their prevalence provides evidence of a declining standard of living in many regions. Profound differences exist in the causes of human death between developed and developing regions of the world. Communicable, maternal, and/or prenatal diseases account for 40% of the deaths in developing regions, compared with only 5% in developed regions.[139] While there is a complex set of factors responsible, inadequate food and contaminated water and soil are the major contributors to diseases and other health problems, especially in developing countries.[140]

Disease and malnutrition are interrelated and, as might be expected, parasitic infections and malnutrition coexist where poverty and poor sanitation are endemic. Poverty and lack of sanitation are as severe in certain urban sectors as in rural areas; several studies point to inequalities even within different parts of cities.[141] Urban environments, especially those without proper sanitation, are becoming a cause for concern due to their high potential for the spread of disease due to overcrowding. The increased density of people in urban environments, especially those with poor sanitation, provides no protection from pollution caused by accumulation of city wastes in water, air, and soil, and indeed creates favorable conditions for the rapid spread of infectious diseases that can easily reach epidemic proportions. Indeed, in part due to malnutrition and overcrowding of the world population, several diseases are rapidly increasing. For instance, an estimated 2 billion people are infected with TB and 2.4 billion are infected with malaria.[142]

About 90% of the diseases occurring in developing countries are associated with the lack of clean water; worldwide, about 4 billion cases of disease are contracted from water.[143] Schistosomiasis and malaria, common diseases throughout the tropics, are examples of parasitic diseases associated with aquatic systems. Parasitic hookworms thrive in contaminated moist soils of the tropics and infect about 2 billion people. Approximately 50 million deaths are caused by all diseases from water, food, air, and soil each year.[144]

Intestinal parasites introduced into humans through contaminated food, water, and soil impact health by reducing the nutritional status of infected individuals in various ways. This includes the rapid loss of food nutrients through diarrhea or dysentery, impairment of nutrient absorption, alteration of appetite and food intake, and also blood loss. The above-mentioned hookworms can remove up to 30 cc of blood from a person in a single day, leaving the person weak and susceptible to additional diseases. Estimates are that from 5% to 20% of an afflicted person's daily food intake is metabolized to offset other illnesses and physical stresses caused by disease, thereby diminishing the nutritional status and ability to resist other diseases.[145]

Certainly the nutrition of the world would be improved with a more equitable distribution of the total world food produced. For instance, it might be possible to feed the current 6.1 billion people a minimal but nutritionally adequate diet if all food produced in the world was shared and distributed equally.[146] However, there are problems with this proposal. For example, how many people in developed and developing countries who have more than their basic needs of food resources would be willing to share their food and pay not only for its production, but for its worldwide distribution? Also, if the world population doubles to 12 billion in just the next 54 years, then this option would no longer be possible because of severe shortages of land, water, energy, and biological resources.[147]

Transition to a 2 Billion Optimum Population

The current human population has enormous momentum for continued rapid growth because of the young age distribution in the world population. If the whole world agreed on and adopted a policy so that only about two children were born per couple, about 70 years would pass before the world population finally stabilized at approximately 12 billion[148,149] (Figure 3). On the other hand, a population policy ensuring that each couple produces an average of only 1.5 children would be necessary to achieve the goal of reducing the world population from the current 6.3 billion to an optimal population of approximately 2 billion.[150] If this policy were implemented, more than 100 years would be required to make the adjustment to 2 billion people. Again, the prime difficulty in making the adjustment is the young age distribution and growth momentum of the current world population.

Our suggested 2 billion population carrying capacity for the Earth is based on a European standard of living and sustainable use of natural resources. For land resources, we suggest 0.5 ha of cropland per capita with an intense agricultural production system (about 8 million kcal/ha) to produce a diverse plant and animal diet. In addition, approximately 1 ha of land would be required per capita for a renewable energy system. Of course, it would also be essential to stop all current land degradation associated with soil erosion and other factors. Technologies are currently available for soil conservation in agricultural and forest production; they only need to be implemented.

A reduction in the world population to approximately 2 billion, in addition to a reduced per capita consumption rate, would help reduce water pollution and the current severe pressure on surface and groundwater resources, especially in countries where water shortages will only intensify with population growth.[151,152] If water shortage and pollution problems were reduced, agricultural production would improve and degradation of aquatic ecosystems would decline. If pollution were controlled in most major river and lake systems worldwide, increased fish production would be possible and extinctions of fish and other valuable aquatic species would be limited.

The adjustment of the world population from 6.3 billion to 2 billion could be made over approximately a century if the majority of the people of the world agree that protecting human health and welfare is vital, and all are willing to work to provide a stable quality of life for ourselves and our children. Although a rapid reduction in population numbers to 2 billion humans could cause social, economic, and political problems, the continued rapid growth to 10 or 12 billion people will result in an even more dire situation with potentially greater problems.

Conclusion

During the past century, humans have used the Earth's environmental resources with little or no concern about their continued availability. Meanwhile, human numbers escalate, necessitating greater use of all these resources. Clear evidence presented in this article documents the serious status of these vital resources that are needed to support human life.

Worldwide, balancing the population-resource equation will be difficult because overpopulation, uneven distribution of resources, and environmental degradation have reached unsustainable levels. The impact of the daily addition of a quarter-million people to the world goes unnoticed. Yet, this growth in human numbers is relentless and its adverse affect on the Earth's resources mounts each day and year. In two recent reports, the World Health Organization indicates that more than 3 billion people (half the world population) now are malnourished. This is the largest number and proportion of malnourished humans ever recorded.

Malnourishment is a serious disease, for it increases the human susceptibility to other diseases like TB, malaria, and AIDS. Sick and diseased people find it difficult to work, participate in life's activities, and enjoy their daily lives. Per capita shortages of basic food resources are responsible for much of this malnutrition. Poverty, as well as inadequate or unfair distribution of food supplies, also contributes.

Cereal grains are the mainstay of human diets, comprising 80% of the world food supply. Yet food availability per capita as measured by cereals has been declining since 1984. Meanwhile, the world population is projected to increase from the current 6.2 billion to 12 billion in about 50 years, based on the current growth rate. Even if a policy of two children per couple were adopted tomorrow instead of the current norm of nearly three children, the world population will continue to increase until 2070 before stabilizing at more than 12 billion.

Fifty-eight academies of science, including the U.S. National Academy of Sciences, summed up the problem as follows: "Humanity is approaching a crisis point with respect to the interlocking issues of population, natural resources, and sustainability".[153] Historically, decisions to protect the environment have been based on isolated problems, and usually are made only when catastrophes strike. Instead of examining the problem in a holistic manner, such *ad hoc* decisions protect and/or promote a particular resource or aspect of human well-being, but only in the short-term. Based on past experience, urgent issues concerning human carrying capacity of the world are not being addressed, but rather ignored until the situation becomes intolerable or possibly irreversible.

Despite all projections about human population growth, no one really knows exactly how large the human population will be in 50 years. We know the 6.2 billion people already on Earth are stressing the Earth's land, water, and biological resources and polluting the environment. We know that more than 3 billion malnourished people are too many.

With a democratically determined population control policy that respects basic individual rights, coupled with sound resource conservation policies plus the support of science and technology to enhance energy supplies and protect the integrity of the environment, an optimum population of 2 billion for the Earth can be achieved. With a concerted effort by everyone, the well-being of future generations can be secured within the 21st century. Then most individuals will be free from poverty and starvation and be able to live in an environment capable of sustaining their lives with dignity. If the human population continues to increase and exhaust the Earth's natural resources, nature will control our numbers by disease, hunger, malnutrition, and violent conflicts over resources. The difficult decisions are ours to be made to prevent the imbalance between human numbers and food security from further escalating.

Notes

1   SSBPRC. 1996. *The Yearbook of Energy Statistics of China in 1996*. Beijing, China: State Statistical Bureau of PRC, Chinese Statistical Press.

2   PRB. 2001. *World Population Data Sheet*. Washington, DC: Population Reference Bureau.

3   *Ibid.*

4   *Ibid.*

5   Weeks, J.R. 1986. *Population: An Introduction to Concepts and Issues*. Third Edition. Belmont, CA: Wadsworth Publishing Co.

6   Population Action International. 1993. *Challenging the Planet: Connections Between Population and Environment*. Washington, DC: Population Action International.

7   Pimentel D., O. Bailey, P. Kim, E. Mullaney, J. Calabrese, F. Walman, F. Nelson, & X. Yao. 1999. Will the limits of the Earth's resources control human populations? *Environment, Development and Sustainability*, 1: 19–39.

8   FAO. 1961–1999. *Quarterly Bulletin of Statistics*. Food and Agriculture Organization of the United Nations.

9   PRB. 2001. *World Population Data Sheet*. Washington, DC: Population Reference Bureau.

10   WHO. 1996. *Micronutrient Malnutrition—Half of the World's Population Affected* (Pages 1–4 No. Press Release WHO No. 78). World Health Organization.

11   WHO 2000. *Malnutrition Worldwide*. http://www.who.int/nut/malnutrition_worldwide.htm, accessed July 27, 2000.

12   Sommer, A. & K.P. West. 1996. *Vitamin A Deficiency: Health, Survival and Vision*. New York: Oxford University Press.

13 Tomashek, K.M., B.A. Woodruff, C.A. Gotway, P. Bloand, & G. Mbaruku. 2001. Randomized intervention study comparing several regimens for the treatment of moderate anemia refugee children in Kigoma region, Tanzania. *American Journal of Tropical Medicine and Hygiene* 64 (3/4): 164–171.

14 FAO. 1991. *Food Balance Sheet.* Food and Agriculture Organization of the United Nations.

15 FAO. 1998. *Food Balance Sheet.* Food and Agriculture Organization of the United Nations.

16 Pimentel, D. & M. Pimentel. 1996. *Food, Energy and Society.* Boulder, CO: Colorado University Press.

17 Buringh, P. 1989. Availability of agricultural land for crop and livestock production. In: *Food and Natural Resources,* eds. D. Pimentel and C.W. Hall. San Diego: Academic Press.

18 FAOSTAT. 1994. *Statistical Database.* Food and Agricultural Organization of the United Nations.

19 Lal, R. 1989. Land degradation and its impact on food and other resources. In: *Food and Natural Resources,* ed. D. Pimentel. San Diego: Academic Press.

20 Giampietro, M. and D. Pimentel. 1994. Energy utilization. In: *Encyclopedia of Agricultural Science,* eds. C.J. Arntzen and E.M. Ritter. San Diego, CA: Academic Press.

21 Doeoes, B.R. 1994. Environmental degradation, global food production, and risk for larger-scale migrations. *Ambio* 23 (2): 124–130.

22 USBC. 2000. *Statistical Abstract of the United States 2000.* Vol. 200th ed. Washington, DC: U.S. Bureau of the Census, U.S. Government Printing Office.

23 Leach, G. 1995. *Global Land and Food in the 21st Century.* Stockholm: International Institute for Environmental Technology and Management.

24 Brown, L.R. 1997. *The Agricultural Link.* Washington, DC: Worldwatch Institute.

25 Wen, D. & D. Pimentel. 1992. Ecological resource management to achieve a productive, sustainable agricultural system in northeast China. *Ecosystems and Environment* 41: 215–230.

26 Alexandratos, N. 1995. *World Agriculture: Towards 2010.* Rome: Food and Agriculture Organization of the United Nations and John Wiley & Sons.

27 Lal, R. & B.A. Stewart. 1990. *Soil Degradation.* New York: Springer-Verlag.

28 Pimentel, D. & N. Kounang. 1998. Ecology of soil erosion in ecosystems. *Ecosystems* (1998) 1: 416–426.

29 Lal, R. & B.A. Stewart. *Op cit.*

30 Pimentel, D., C. Harvey, P. Resosudarmo, K. Sinclair, D. Kurz, M.

McNair, S. Crist, L. Shpritz, L. Fitton, R. Saffouri, & R. Blair. 1995. Environmental and economic costs of soil erosion and conservation benefits. *Science* 267: 1117–1123.

31 Houghton, R.A. 1994. The worldwide extent of land-use change. *BioScience*. 44(5): 305–313.

32 Pimentel, D., et. al. *Op. cit.*

33 Houghton, R.A. *Op. cit.*

34 WRI. 1996. *World Resources 1996–97.* Washington, DC: World Resources Institute.

35 Myers, N. 1990. *The Nontimber Values of Tropical Forests.* Forestry for sustainable development program. University of Minnesota. November, 1990. Report 10.

36 Lal, R. & F.J. Pierce. 1991. *Soil Management for Sustainability.* Ankeny, Iowa: Soil and Water Conservation Soc. in Coop. with World Assoc. of Soil and Water Conservation and Soil Sci. Soc. of Amer.

37 Pimentel, D., C. Harvey, P. Resosudarmo, K. Sinclair, D. Kurz, M. McNair, S. Crist, L. Shpritz, L. Fitton, R. Saffouri, & R. Blair. 1995. Environmental and economic costs of soil erosion and conservation benefits. *Science* 267: 1117–1123.

38 Kendall, H.W. & D. Pimentel. 1994. Constraints on the expansion of the global food supply. *Ambio* 23: 198–205.

39 OTA. 1982. *Impacts of Technology on U.S. Cropland and Rangeland Productivity.* Washington, DC: Office of Technology, U.S. Congress.

40 Elwell, H.A. 1985. An assessment of soil erosion in Zimbabwe. *Zimbabwe Science News* 19: 27–31.

41 Troeh, F.R, J.A Hobbs, & R.L. Donahue. 1991. *Soil and Water Conservation.* 2nd ed., Englewood Cliffs, NJ: Prentice Hall.

42 Pimentel, D., C. Harvey, P. Resosudarmo, K. Sinclair, D. Kurz, M. McNair, S. Crist, L. Shpritz, L. Fitton, R. Saffouri, & R. Blair. 1995. Environmental and economic costs of soil erosion and conservation benefits. *Science* 267: 1117–1123.

43 Wen, D. 1993. Soil erosion and conservation in China. In: *Soil Erosion and Conservation,* ed. D. Pimentel. New York: Cambridge University Press.

44 McLaughlin, L. 1993. A case study in Dingxi County, Gansu Province, China. *World Soil Erosion and Conservation.* ed. D. Pimentel. Cambridge: Cambridge University Press.

45 Tolba, M.K. 1989. Our biological heritage under siege. *BioScience* 39: 725–728.

46 Pimentel, D. 1993. *World Soil Erosion and Conservation.* Cambridge: Cambridge University Press.

47 Gupta, J. P., & Raina, P. 1996. Wind erosion and its control in hot arid areas of Rajasthan. In: *Wind Erosion in West Africa: The Problem and its Control.* eds. B. Buerkert, B. E. Allison, & M. von Oppen. Berlin: Margraf Verlag.

48   Parrington, J.R., W.H. Zoller, & N.K. Aras. 1983. Asian dust: seasonal transport to the Hawaiian Islands. *Science* 246: 195–197.

49   Pimentel, D. & N. Kounang. 1998. Ecology of soil erosion in ecosystems. *Ecosystems* (1998) 1: 416–426.

50   Zhang, X, D.E. Walling, T.A. Quine, & A. Wenn. 1997. Use of reservoir deposits and caesium-137 measurements to investigate the erosional response of a small drainage basin in the rolling loess plateau region of China. *Land Degradation and Development.* 8: 1–16.

51   Myers, N. 1993. *Gaia: An Atlas of Planet Management.* Garden City, NY: Anchor Press/Doubleday & Co. Inc.

52   Pimentel, D., C. Harvey, P. Resosudarmo, K. Sinclair, D. Kurz, M. McNair, S. Crist, L. Shpritz, L. Fitton, R. Saffouri, & R. Blair. 1995. Environmental and economic costs of soil erosion and conservation benefits. *Science* 267: 1117–1123.

53   Gleick, P.H. 1996. Basic water requirements for human activities: meeting basic needs. *Water International* 21(2): 83–92.

54   Postel, S. 1996. *Dividing the Waters: Food Security, Ecosystem Health, and the New Politics of Scarcity.* Vol. 132. Washington, DC: Worldwatch Institute.

55   Hinrichsen, D. 1998. Feeding a future world. *People and the Planet* 7 (1): 6–9.

56   Gleick, P.H. 1993. *Water in Crisis.* New York: Oxford University Press.

57   WRI. 1994. *World Resources* 1994–95. Washington, DC: World Resources Institute.

58   Brown, L.R. 1995. *Who Will Feed China? Wake-Up Call for a Small Planet.* New York: W.W. Norton and Co.

59   Pimentel, D., J. Houser, E. Preiss, O. White, H. Fang, L. Mesnick, T. Barsky, S. Tariche, J. Schreck, & S. Alpert. 1997. Water resources: agriculture, the environment, and society. *BioScience* 47 (2): 97–106.

60   Sheridan, D. 1983. The Colorado—an engineering wonder without enough water. *Smithsonian.* February: 62; 45–54.

61   UNEP. 1991. *Freshwater Pollution.* Global Environment Monitoring System, Nairobi, Kenya: United Nations Environment Programme.

62   Covich, A.P. 1993. Water and ecosystems. In *Water in Crisis,* ed. P.H. Gleick. New York: Oxford University Press.

63   Postel, S. 1989. Water for agriculture: facing the limits. *Worldwatch Paper* 93. Washington, DC: Worldwatch Institute.

64   UNFPA. 1991. *Population and the Environment: The Challenges Ahead.* New York: United Nations Fund for Population Activities, United Nations Population Fund.

65   Postel, S. 1997. *Last Oasis: Facing Water Scarcity.* New York: W.W. Norton and Co.

66   USWRC. 1979. *The Nation's Water Resources.* 1975–2000. Vol. 1–4. Second National Water Assessment, Washington, DC: U.S. Water Resources Council.

67 Beaumont, P. 1985. Irrigated agriculture and groundwater mining on the high plains of Texas. *Environmental Conservation* 12: 11.

68 Soule, J.D. & D. Piper. 1992. *Farming in Nature's Image: An Ecological Approach to Agriculture.* Washington, DC: Island Press.

69 Gleick, P.H. 1993. *Water in Crisis.* New York: Oxford University Press.

70 Postel, S. 1997. *Last Oasis: Facing Water Scarcity.* New York: W.W. Norton and Co.

71 White, R. 2001. *Evacuation of Sediments From Reservoirs.* Bristol, UK: Thomas Telford, Limited.

72 Postel, S. *Op. cit.*

73 Pimentel, D., J. Houser, E. Preiss, O. White, H. Fang, L. Mesnick, T. Barsky, S. Tariche, J. Schreck, & S. Alpert. 1997. Water resources: agriculture, the environment, and society. *BioScience* 47 (2): 97–106.

74 Pimentel, D., M. Herz, M. Whitecraft, M. Zimmerman, R. Allen, K. Becker, J. Evans, B. Hussan, R. Sarsfeld, A. Grosfeld, & T. Seidel. 2002. Renewable energy: energetic, economic, and environmental issues. *BioScience* 52(12): 1111–1120.

75 Postel, S. 1997. *Last Oasis: Facing Water Scarcity.* New York: W.W. Norton and Co.

76 EPA. 1976. *Evaluating Economic Impacts of Programs for Control of Saline Irrigation Return Flows: A Case Study of the Grand Valley, Colorado.* Denver, Colorado: U.S. Environmental Protection Agency.

77 Pimentel, D., J. Houser, E. Preiss, O. White, H. Fang, L. Mesnick, T. Barsky, S. Tariche, J. Schreck, & S. Alpert. 1997. Water resources: agriculture, the environment, and society. *BioScience* 47 (2): 97–106.

78 Pimentel, D., M. Herz, M. Whitecraft, M. Zimmerman, R. Allen, K. Becker, J. Evans, B. Hussan, R. Sarsfeld, A. Grosfeld, & T. Seidel. 2002. Renewable energy: energetic, economic, and environmental issues. *BioScience* 52(12):1111–1120.

79 USBC. 2000. *Statistical Abstract of the United States 2000.* Vol. 200th ed. Washington, DC: U.S. Bureau of the Census, U.S. Government Printing Office.

80 WHO. 1992. *Our Planet, Our Health: Report of the WHO Commission on Health and Environment.* Geneva: World Health Organization.

81 NGS. 1995. *Water: A Story of Hope.* Washington, DC: National Geographic Society.

82 EPA. 1994. *Quality of Our Nation's Water 1994.* Washington, DC: U.S. Environmental Protection Agency.

83 Nash, L. 1993. Water quality and health. In: *Water In Crisis: A Guide to the World's Fresh Water Resources,* ed. P. Gleick. Oxford: Oxford University Press.

84 UNEP. 2001. Silt behind dams worsens water shortages. Accessed at http://news.bbc.co.uk/hi/english/sci/tech/newsid1691000/169172.stm. (May 24, 2002).

85 White, R. 2001. *Evacuation of Sediments From Reservoirs*. Bristol, UK: Thomas Telford, Limited.

86 Pimentel, D. & M. Pimentel. 1996. *Food, Energy and Society*. Boulder, CO: Colorado University Press.

87 *International Energy Annual, 1995*. DOE/EIA-0219[95]. U.S. Department of Energy, Washington, DC.

88 PRB. 1996. *World Population Data Sheet*. Washington, DC: Population Reference Bureau.

89 *International Energy Annual, 1995*. DOE/EIA-0219[95]. U.S. Department of Energy, Washington, DC.

90 PRB. 2001. *World Population Data Sheet*. Washington, DC: Population Reference Bureau.

91 *International Energy Annual, 2001*. DOE/EIA. U.S. Department of Energy, Washington, DC.

92 Pimentel, D., M. Herz, M. Whitecraft, M. Zimmerman, R. Allen, K. Becker, J. Evans, B. Hussan, R. Sarsfeld, A. Grosfeld, & T. Seidel. 2002. Renewable energy: energetic, economic, and environmental issues. *BioScience* 52(12): 1111–1120.

93 *International Energy Annual, 2001*. DOE/EIA. U.S. Department of Energy, Washington, DC.

94 USBC. 2000. *Statistical Abstract of the United States 2000*. Vol. 200th ed. Washington, DC: U.S. Bureau of the Census, U.S. Government Printing Office.

95 Pimentel, D. & M. Pimentel. 1996. *Food, Energy and Society*. Boulder, CO: Colorado University Press.

96 DOE. 1991. *Annual Energy Outlook With Projections to 2010*. Washington, DC: U.S. Department of Energy, Energy Information Administration.

97 DOE. 1995. *Annual Energy Outlook With Projections to 2010*. Washington, DC: EIA, USDOE.

98 Smil, V. 1984. *The Bad Earth: Environmental Degradation in China*. Armonk, NY: M.E. Sharpe, Inc.

99 Wen, D. & D. Pimentel. 1992. Ecological resource management to achieve a productive, sustainable agricultural system in northeast China. *Ecosystems and Environment* 41: 215–230.

100 *International Energy Annual, 2001*. DOE/EIA. U.S. Department of Energy, Washington, DC.

101 Wen, D. & D. Pimentel. *Op. cit.*

102 *Vital Signs 2001*. The Worldwatch Institute. Washington, DC.

103 Ivanhoe, L.F. 1995. Future world oil supplies: there is a finite limit. *World Oil*. October: 77–88.

104 Campbell, C.J. 1997. *The Coming Oil Crisis*. New York: Multi-Science Publishing Company & Petroconsultants S.A.

105 Duncan, R.C. 1997. *The World Petroleum Life-Cycle: Encircling the Production Peak*. Space Studies Institute May 9: 1–8.

106 Youngquist, W. 1997. *Geodestinies: The Inevitable Control of Earth Resources Over Nations and Individuals.* Portland, OR: National Book Company.

107 BP. 2001. *British Petroleum Statistical Review of World Energy.* London: British Petroleum Corporate Communications Services.

108 Bartlett, A.A. & R.A. Ristinen. 1995. Natural gas and transportation. *Physics and Society* 24 (4): 9–10.

109 Youngquist, W. 1997. *Geodestinies: The Inevitable Control of Earth Resources Over Nations and Individuals.* Portland, OR: National Book Company.

110 Campbell, C.J. 1997. *The Coming Oil Crisis.* New York: Multi-Science Publishing Company & Petroconsultants S.A.

111 Youngquist, W. *Op. cit.*

112 *Ibid.*

113 USBC. 2000. *Statistical Abstract of the United States 2000.* Vol. 200th ed. Washington, DC: U.S. Bureau of the Census, U.S. Government Printing Office.

114 Youngquist, W. Personal communication. Eugene, Oregon, 2002.

115 Pimentel, D., M. Herz, M. Whitecraft, M. Zimmerman, R. Allen, K. Becker, J. Evans, B. Hussan, R. Sarsfeld, A. Grosfeld, & T. Seidel. 2002. Renewable energy: energetic, economic, and environmental issues. *BioScience* 52(12):1111–1120.

116 Pimentel, D., G. Rodrigues, T. Wang, R. Abrams, K. Goldberg, H. Staecker, E. Ma, L. Brueckner, L. Trovato, C. Chow, U. Govindarajulu, & S. Boerke. 1994. Renewable energy: economic and environmental issues. *BioScience* 44: 536–547.

117 *Ibid.*

118 Pimentel, D., U. Stachow, D.A. Takacs, H.W. Brubaker, A.R. Dumas, J.J. Meaney, J. O'Neil, D.E. Onsi, & D.B. Corzilius. 1992. Conserving biological diversity in agricultural/forestry systems. *BioScience* 42: 354–362.

119 Sagoff, M. 1995. Carrying capacity and ecological economics. *BioScience* 45(9): 610–620.

120 Wilson, E. O. 1988. *Biodiversity.* Washington, DC: National Academy of Sciences.

121 Vietmeyer, N. 1995. Applying biodiversity. *Journal of the Federation of American Scientists* 48 (4): 1–8.

122 Reid, W.V. & K.R. Miller. 1989. *Keeping Options Alive: The Scientific Basis for Conserving Biodiversity.* Washington, DC: World Resources Institute.

123 Kellert, R.S. & E.O. Wilson. 1993. *The Biophilia Hypothesis.* Washington, D.C.: Island Press.

124 O'Toole, C. 1993. Diversity of native bees and agroecosystems. In: *Hymenoptera and Biodiversity,* eds. J. LaSalle & I.D. Gault. Wallingford, Oxon, U.K.: CAB International.

125 Pimentel, D., C. Wilson, C. McCullum, R. Huang, P. Dwen, J. Flack, Q. Tran, T. Saltman, & B. Cliff. 1997. Economic and environmental benefits of biodiversity. *BioScience* 47 (11): 747–758.

126 DeBach, P. & D. Rosen. 1991. *Biological Control by Natural Enemies.* New York: Cambridge University Press.

127 Pimentel, D. 1997. *Techniques for Reducing Pesticides: Environmental and Economic Benefits.* Chichester, UK: John Wiley & Sons.

128 Pimentel, D., L. Lach, R. Zuniga, & D. Morrison. 2000. Environmental and economic costs of nonindigenous species in the United States. *BioScience* 50(1), 53–65.

129 Pimentel, D., S. McNair, J. Janecka, J. Wightman, C. Simmons, C. O'Connell, E. Wong, L. Russel, J. Zern, T. Aquino, & T. Tsomondo. 2001. Economic and environmental threat of alien plant, animal and microbe invasions. *Agriculture, Ecosystems and the Environment,* 84, 1–20.

130 Pimentel, D., et. al. *Op cit.*

131 Pimentel, D. Unpublished data.

132 Pimentel, D. 1997. *Techniques for Reducing Pesticides: Environmental and Economic Benefits.* Chichester, UK: John Wiley & Sons.

133 Reid, W.V. & K.R. Miller. 1989. *Keeping Options Alive: The Scientific Basis for Conserving Biodiversity.* Washington, DC: World Resources Institute.

134 Western, D. 1989. Conservation without parks: Wildlife in rural landscape. In: *Conservation for the Twenty-first Century,* eds. D. Western & M.C. Pearl. New York: Oxford University Press.

135 Pimentel, D., U. Stachow, D.A. Takacs, H.W. Brubaker, A.R. Dumas, J.J. Meaney, J. O'Neil, D.E. Onsi, & D.B. Corzilius. 1992. Conserving biological diversity in agricultural/forestry systems. *BioScience* 42: 354–362.

136 WHO. 1992. *Our Planet, Our Health: Report of the WHO Commission on Health and Environment.* Geneva: World Health Organization.

137 WHO. 1995. *Bridging the Gaps.* Geneva: World Health Organization.

138 Murray, C.J.L. & A.D. Lopez. 1996. *The Global Burden of Disease: A Comprehensive Assessment of Mortality and Disability from Diseases, Injuries, and Risk Factors In 1990 and Projected to 2020.* Cambridge: Harvard School of Public Health.

139 WHO. 1994. *Global Comparative Assessments in the Health Sector: Disease Burden, Expenditures and Intervention Packages.* Geneva: World Health Organization.

140 Pimentel, D., M. Tort, L. D'Anna, A. Krawic, J. Berger, J. Rossman, F. Mugo, N. Doon, M. Shriberg, E.S. Howard, S. Lee, & J. Talbot. 1998. Increasing disease incidence: environmental degradation and population growth. *BioScience* 48(10): 817–826.

141 *Ibid.*

142 WHO. 1998. *WHO Information Fact Sheet.* Geneva: World Health Organization.

143 WHO. 1992. *Our Planet, Our Health: Report of the WHO Commission on Health and Environment.* Geneva: World Health Organization.

144 WHO. 1995. *Bridging the Gaps.* Geneva: World Health Organization.

145 Pimentel, D. & M. Pimentel. 1996. *Food, Energy and Society.* Boulder, CO: Colorado University Press.

146 Cohen, J.E. 1995. *How Many People Can the Earth Support?* New York: Rockefeller University.

147 Abernethy, V. 1993. *Population Politics: The Choice That Shapes Our Future.* New York: Insight Books.

148 Weeks, J.R. 1986. *Population: An Introduction to Concepts and Issues.* Third Edition. Belmont, CA: Wadsworth Publishing Co.

149 Population Action International. 1993. *Challenging the Planet: Connections Between Population and Environment.* Washington, DC: Population Action International.

150 Pimentel, D., R. Harman, M. Pacenza, J. Pecarsky, & M. Pimentel. 1994. Natural resources and an optimum human population. *Population and Environment* 15: 347–369.

151 Postel, S. 1997. *Last Oasis: Facing Water Scarcity.* New York: W.W. Norton and Co.

152 Pimentel, D., J. Houser, E. Preiss, O. White, H. Fang, L. Mesnick, T. Barsky, S. Tariche, J. Schreck, & S. Alpert. 1997. Water resources: Agriculture, the environment, and society. *BioScience* 47 (2): 97–106.

153 NAS. 1994. *Population Summit of the World's Scientific Academies.* Washington, DC: National Academy of Sciences Press.

4

Survival from the Brain's Perspective

Walter Lowen

It seems to me that it is often very difficult to get people to help themselves, no matter how obvious the benefit. There are many examples. I'll cite one:

I'm reminded of a real situation when, years ago, I was active in Volunteers for International Technical Assistance (VITA). It involved an effort to increase food production in a poor Asian country (I've forgotten the details). A farmer whose fields were easily visible from the whole valley was selected as a demonstration project of what could be achieved with modern methods of crop rotation, contour plowing, judicious use of fertilizer, etc. The assumption was that all the other farmers would seek him out to inquire how he achieved such superior crops, and thereby he could educate a whole region.

What actually happened was that the demonstration farmer selected found himself, initially, in oodles of meetings with region-al "bigwigs." Bigwigs wore their pants down to their ankles, whereas farmers wore them just below the knee, which was more practical for farmers. The farmer selected for the demonstration project viewed pant length as a status symbol and gradually adopted the longer style of his administrative cohorts with whom he spent so much time. But when it came time to spread the word, the other farmers, who had no use for or desire to talk with bigwigs, avoided the demonstration farmer—now regarded as a bigwig himself—and so the whole experiment failed.

This true anecdote is an example of the brain's egocentricity. Clearly it looks out for the best interests of the master it serves, which is another way of stating the brain's basic role—to help the organism survive. It does not exist to help the environment survive; no, in man it has a much more complex job. This complexity is the price evolution had to pay to get beyond the earlier, hard-wired, simple stimulus-response life forms exemplified by the moth. In mammals, and certainly in man, new adaptations lead to far greater complexity.

One of the aspects at play in the unfortunate outcome of the anecdote cited earlier is a complexity characterized as a "hierarchy of needs," which Claire Graves studied in job settings.[1] People whose most basic needs (such as food and shelter) are not yet fully satisfied can be induced to work harder merely by the promise of more money, as illustrated by the invention of piecework. But once such basic needs are met, other more abstract needs become the currency of value, such as security, titles, the corner office, or the personal parking space. People have quit their jobs if such status needs were violated.

With so much of the world still "underdeveloped," (the politically incorrect but accurate pre-euphemism), much of the world is still too low in its hierarchy of needs to be motivated by concerns of global warming. If, collectively, we need to address the future, perhaps our most pressing need is to eliminate poverty and suffering throughout the world. But even if we could agree on that goal, it would not ensure that people as a whole would make the well-being of the external environment a priority concern. The difficulty is built into our brains.

As I suggested in my book, *Dichotomies of the Mind,* the organization of the brain, and therefore also the mind, is dichotomous.[2] The human brain is bifurcated in its interests by the fact that it must deal with not only one, but two environments. One is the internal environment (i), which is everything within the skin, also referred to as "the self." Significantly, this i-environment includes the brain itself. The other environment is the external environment (e), which is everything outside the skin, also referred to as the "real world." These two environments (i and e) form an opposition pair, and the brain deals with each in very different ways. This opposition stems from the fact that for the self (i) to survive, it is both dependent upon and threatened by the external environment (e). In short, "eat or be eaten."

The dichotomous challenges are reflected in a dichotomous organization of the cerebral cortex.[3] In man, the cerebral cortex has a deep fold which runs roughly from one ear to the other over the top of the cortex. It is called the *central fissure*, dividing the cortex into a front half and a back half. The front monitors the internal environment for slight deviations from homeostasis, interprets such small differences as an indication of needs, and then busies itself with developing action plans to meet those needs. Ultimately it takes actions by directing instructions to the motor cortex, lying adjacent to and in front of the central fissure. The resulting actions, designed to maintain a stable i-environment, may bring about changes in the e-environment.

Concurrently and contrastingly, the back half of the cerebral cortex receives a stream of information about the external environment gathered by the senses. It uses that information, together with relevant past experiences retrieved from long-term memories also located in the back half, to build a model of the real world. It monitors these real world models for significant changes, indicating either danger or opportunity.*

This front/back or e/i dichotomy—only briefly outlined above but more fully discussed in Kent (1981), Lowen (1982), and Lowen & Klir (1989)[4, 5, 6]—suggests to me that not only is there a built-in bias favoring i over e, but unless the status of e reaches proportions which make it a threat to i, little change in attitude toward impending dangers can be expected.

Another bifurcation which might be labeled the up/down dichotomy deals with interactions between the cerebral cortexes and lower-lying brains, a section of the organ much older in evolutionary time. The evolution of the cerebral cortex, in mammals

---

* I am indebted to Prof. Bert Horwitz of SUNY-Binghamton for the fascinating insight provided by his knowledge of Chinese: Wei[(1)] Xian[(3)] (pronounced "way seh en") is the Chinese word for danger. Ji[(1)] Hui[(4)] (pronounced "gee khway) is the Chinese word for opportunity. Compounding them as wei[(1)] ji[(1)] (pronounced "way gee") a new word meaning "crisis" conveys the idea that a crisis presents both a danger and, more important, an opportunity for change, implying that some crises should be welcome. (The numbers after the characters identify the tone number, a total of five in Chinese.)

especially, created an organism which could respond to challenges posed by the external environment with greater flexibility. Unfortunately the front half of the cerebral cortex, especially the pre-frontal cortex, relies heavily on lower brain structures such as the limbic system to evaluate the potential effectiveness of contemplated actions. Since older brains counted on unchanging environments, this promising flexibility for survival could be neutralized in a drastically changing world.

There may, however, be some hope in what may be called the right/left brain dichotomy. Gazzaniga, one of the early participants in split-brain research, refers to a unique capability in the left brain hemisphere which he calls the *interpreter*.[7] The interpreter can "cook up" answers to overly complex issues, "looking for order and reason even where they don't exist." I believe it is the interpreter which makes belief systems possible and causes them to seem so real.[8]

Belief systems can influence behavior in powerful ways. If we consider how native Americans instilled in their children a reverence for preserving the external environment, one could imagine that Survival Research might be directed toward early childhood education with the aim of exploiting the inventive power of the interpreter toward belief systems practiced by the earliest Americans.

This look at how the future might develop as seen from the perspective of the human brain totally ignores the additional complexities created by the various dispositions of consciousness. Even though we do not yet understand the unconscious, Jung alerted us to the added complexities created by personality differences, a consequence of the various dispositions of consciousness.[9] As is so often the case, when complexity is high, it will be something totally unexpected that is most likely to determine the future.

Notes

1 Graves, Clare W. 1966. Deterioration of work standards. *Harvard Business Review* 44(5); Sept.–October.

2 Lowen, Walter. 1982. *Dichotomies of the Mind.* New York: John Wiley & Sons.

3 Kent, Ernest W. 1981. *The Brains of Men and Machines.* Peterborough, N.H.: Byte/McGraw Hill.

4 *Ibid.*

5 Lowen, Walter. 1982. *Dichotomies of the Mind.* New York: John Wiley & Sons.

6 Lowen, Walter & Geroge J. Klir. 1989. The world of mathematics mirrors the organization of the mind. (part II, *Mathematics, the Mind, and the Brain*). presented at Edinburgh University. To be republished in the *International Journal of General Systems.*

7 Gazzaniga, Michael S. 1998. *The Mind's Past.* Berkeley: University of California Press.

8 Lowen, Walter. 2002. The gift of life. *International Journal of General Systems* 31(4): pp 431–433.

9 Jung, C. G. 1976. *Psychological Types.* Princeton, N.J.: Princeton University Press.

5

Biology is Destiny
Only if We Ignore It

Jerome H. Barkow

It is deeply satisfying to point out how human stupidity, greed, carelessness, and irresponsibility are the real causes of our problems. Countless popular books and newspaper columns and editorials have had at their core a Victorian moral earnestness combined with astonishment over the folly of our fellows. But there is another Victorian discourse that, ultimately, is both more satisfying and potentially more effective than any rhetoric of virtue and its lack: the discourse of Charles Darwin and of human evolution. No doubt, seeking moral causes and cures for today's problems of survivability and sustainability is unavoidable, but perhaps not as productive as one might wish. It is time to try the Darwinian perspective; it presents our problems not as products of moral imperfection, but of an evolved psychology—one better suited for the environments of our ancestors than for those in which we now find ourselves, but which nonetheless underlies our societies and institutions.

Darwin taught us that our species is the product of biological evolution, and that evolution by natural selection is a process of unintelligent design. That is, evolution is a blind process conceptually similar to the blind processes that cause the molecules of crystals to align themselves in perfect order or the pebbles of a stream bed to be sorted by the rushing of water. As the philosopher Daniel Dennett has so lucidly explained, Darwinian evolution is algorithmic, a matter of differential survival and reproduction. Given a surplus of offspring, and given that they differ somewhat from one another, in any specified environment some individuals are more likely to survive and reproduce than are others. Offspring resemble their parents. Whichever variations make individuals slightly more likely to survive and reproduce than others will therefore be more common in the next generation. If the environment that led to selection of these attributes remains reasonably constant then, from generation to generation, these traits will become more frequent in the species and observers may call the accumulating changes "evolution by natural selection."[1] However, the products of this process have at least three unfortunate imperfections.

First, it is apparent that a species is primarily adapted only to past environments; adaptation to the current environment usually depends on the extent to which it resembles that of the past. Evolution looks backward, not forward. Traits do not evolve because they might someday be useful, but only because they are already useful. True, as environments change, so too do selection pressures, and traits may change their function. But this process is always slow. For most of human evolutionary history we lived in small, competing bands of people who gathered and hunted. We did not suddenly settle down to become farmers some 10,000 years ago, as was once believed. We now understand that both before and after that time, where there were stable sources of food such as shellfish or salmon we developed complex sedentary societies that were nevertheless based on hunting and gathering. We also realize now that we domesticated plants well before we became full-time farmers.[2,3] Since the cultivation of crops and domestication of livestock became widespread some 10 to 14 thousand years ago, most of our species has become sedentary. However, given that our species is at least 100,000 years old, it seems very likely that we are better adapted to the older range of past environments than to our more recent agricultural-industrial-urban ways of life. Of course, as will be discussed subsequently, we construct our own environments and many of their features may well reproduce aspects of past environments.[4]

Second, there is no implication here that evolution produces perfect biological machines. An organism does not need to be perfect to survive and reproduce, it needs only to be slightly less poorly adapted to its environment than are other members of its species. At no point in the history of *Homo sapiens,* therefore, is there any reason to believe that we were perfect creatures in perfect harmony either with our environments or with one another. Knowledge of our past may be invaluable for understanding the present and for planning the future, but there is no returning to a golden age that never was.

Third, evolution compromises.[5] For example, genes have multiple effects (pleiotropy), and the final frequency in the gene pool of any particular gene is resultant of forces. Thus, evolutionists explain senescence—why we grow old—as in part due to the likelihood that some of the genes that make us more successful early in life in things like having offspring, later on have other effects that are deleterious.[6] Similarly, large body size may make us better competitors in physical violence and permit us to birth bigger, more mature babies but, when food is in short supply, it puts us at greater risk of malnourishment. Both the rate at which humans age and the size of our bodies are thus evolutionary compromises, not products of a process of perfection.

In recent years, evolutionists have turned their attention to human psychology and society. Various labels have been applied to their efforts—human sociobiology, behavioral ecology, evolutionary psychology, and so forth. But these controversies over nomenclature are minor compared to those involving the legitimacy and even honorability of applying Darwin to human behavior. After all, the 20th and early 21st centuries have seen much evil justified by a pseudo-biology and a false genetics, and the danger of misuse of Darwin by those who seek to rationalize their claims of group exclusivity and superiority remains real. No doubt it is these history-based fears that have permitted the last of the -centrisms, species-centrism, to remain respectable in at least some circles (particularly those of the social sciences).[7]

Ours, however, is an age of biology, a period in which genomics and proteomics have replaced particle physics as the expensive "big science" glamour fields, a time when the brain and consciousness are beginning to be understood and biomedicine looms large in public awareness. Our time is also one in which concern for the environment and biodiversity is growing, and in which much of the public (at least, the reading public) is prepared to accept, finally, that human beings are part of the natural order. People who are vegans for ethical reasons or who take James Lovelock's Gaia hypothesis[8,9] seriously are likely to agree that the Darwinian theories that have proven themselves so powerful for every other animal species must now be applied to our own. The question, of course, is just how they are to be applied.

These are still early days for the Darwinian understanding of human nature. No doubt in the next half century evolutionists will develop a firmly-grounded consensus about our evolved human nature, providing a solid foundation for those concerned with social policy. Unfortunately, questions of survivability and sustainability will not wait 50 years. What follows is some of what I believe the future consensus about human evolved psychology will include. It is, at least, consistent with current literature.

"Those who cannot remember the past are condemned to repeat it," goes George Santayana's famous adage. But the saying only serves to give us false hope; remembering the past and its cargo of grievances is more often likely to provoke than prevent conflict. Of course the study of history is useful, but without an understanding of human nature there is absolutely no evidence that it enables us to avoid the repetition of error—much conflict has been based on past grievances, after all. Combining our study of history with an analysis of our evolved human nature therefore seems worth trying. Human nature, however, covers much ground, and this brief paper must perforce be selective. The aspects of our evolved human nature discussed below were chosen because they illustrate why challenges to sustainability and survivability recur. Other choices could easily have been made.

Warning: no assumption is being made here that awareness of human nature somehow alters it or permits us to transcend it. It is hoped only that knowledge of ourselves will permit the development of social policies that more effectively achieve our shared goals, goals having to do with the sustainability of human endeavors and the survival of human societies.

Ethnocentrism

I use "ethnocentrism" here in a very broad manner, to cover the general tendency for us to form into in-groups and out-groups. The phenomenon has been studied by many social psychologists and other scholars, including a number of evolutionists. Human beings everywhere form bands or cliques or groups whose members see themselves as different from (and in at least some ways, superior to) other bands or cliques or groups. In extreme cases, only members of the in-group are considered "human," while out-group members are dehumanized; the norms of ethical conduct applied to relations with other in-group members are often not applied to outsiders, who may be cheated or even killed. Even when out-group members are considered human, the intrinsic superiority of the in-group is seldom questioned. Anything may count as a marker of group membership and a maintainer of group boundaries: language, accent, hairstyle, religious belief, skin color, food preferences, etc. The peoples of different countries may be ethnocentric, but students in high schools also form themselves into antagonistic membership groups, the different divisions of a large corporation may work to sabotage one another's efforts, and government departments may deliberately withhold vital information from one another. The social causes and consequences of these various forms of ethnocentrism obviously differ, but parsimony suggests that it is very likely that they share the same evolutionary psychology, the same evolved mechanisms.

In modern complex societies, individuals may have numerous affiliations, each of which can potentially serve as a group membership marker. Some affiliations are ascribed by birth—particularly those involving ethnicity—but many others are achieved (e.g., occupational status, educational status, taste in entertainment status, etc.). Politicians and other would-be leaders work to persuade potential followers that the membership they allegedly share is paramount, and all other affiliations either secondary or spurious. Leaders speak in "we" terms: we women, we men, we Canadians, we stamp collectors, we persons living with HIV/AIDS, we members of our labor union, we senior corporate executives, we believers in the true religion, we honest working people, we golfers, and so forth.

In-groups generate solidarity through perceived external threat. Historically, numerous political leaders have promoted or taken advantage of wars in order to provide such threat, and to benefit from the evolved responses of clustering around the leader and the suppression of dissent. Wartime leaders enjoy the kind of support they can only dream of in times of peace (Winston Churchill, immensely popular in Great Britain during World War II, was voted out of office with the return of peace). Metaphorical wars are constantly invoked by leaders in an effort to release ethnocentrism's collective solidarity mechanism, resulting in "wars" against crime, poverty, cancer, etc. The typical seeker of leadership usually makes a generic speech: "We are all members of a great in-group with a great history, and this status should be paramount in our minds; but our in-group is threatened by rival external groups and/or internal factions notable for their moral inferiority and lack of respect for us. Part of our parlous situation is due to the errors of former leaders. But rally around me and I will lead you to establish/re-establish the primacy that is naturally and deservedly ours." Out-groups are generally stereotyped and vilified and perceived by in-group members as much more different from in-group members and more internally coherent, cooperative, and powerful than they in fact are.

Numerous historical and social psychological factors influence how and when the ethnocentrism mechanism is triggered and who benefits from it. The mechanism is a double-edged sword in that it tends to increase within-group cooperation at the cost of fomenting conflict with "rival" groups. But in the absence of strong external threats in-groups tend to dissolve into segments, each of which takes one or more of the others as the rival or threat. From this perspective, world peace would be much easier to achieve if we were all convinced of an extraterrestrial threat (e.g., a comet heading towards Earth, or hostile aliens).

The ethnocentrism response reflects the evolution of our species, which possibly was speeded by a population structure involving small, genetically semi-isolated bands often in conflict with one another. Such a scenario, though somewhat speculative, would explain why the trait would in the past have been adaptive.[10,11] Whatever its origins, however, there is little doubt that the tendency towards the various kinds of ethnocentrism is a pan-human trait, intimately involved both with the constant group conflict and the spates of in-group cooperation that typify our species.

Can an ethnocentrism-prone species survive in an era of cheap bioweaponry and other instruments of mass destruction? Modern transportation and communication permit groups separated by vast distances to perceive each other as threats, while both population mixing due to political and economic migration and the rise of schismatic religious and ideological movements mean that the "other" often lives next door. The moral solution of condemning the more destructive of the ethnocentrisms is necessary, but historically has never been sufficient. Seeking symbolic means of expressing ethnocentrism which do not involve visiting physical or psychological violence upon the other is often suggested and is no doubt also worth doing; however, it should be noted that the Berlin Olympics of 1936 did nothing to prevent World War II. The approach of educating our children to ignore the in-group/out-group distinction has led to only limited success,[12] in part because we human beings create group markers and boundaries much faster than they can be dissolved. However, emphasizing the situational nature of group membership and taking advantage of the multiplicity of in-groups to which we belong might be effective. When in-groups have overlapping memberships and their relative salience is largely situational, it becomes more difficult for would-be leaders to persuade members to act violently against out-groups. Cross-cutting ties helped to contain violence in many pre-colonial African societies, for example.[13]

Even racism, one of the most pernicious forms of ethnocentrism, is not inevitable. While our evolved "cognitive machinery" can lead us to use "race" to classify another as being a member of a competing "coalitional affiliation," cues that signify that the other is indeed a member of our own "alliance" can take precedence. In other words, while our tendency to form in-groups and out-groups may be a deeply embedded aspect of our evolutionary psychology, this does not mean that any particular form of ethnocentrism, including and especially racism, is inescapable.[14,15] That ethnocentrism is part of our evolved human nature does not imply that the recurring calamities associated with it are inevitable. Sociological and educational solutions are always possible, it is only *permanent* solutions that are impossible.

Hierarchizing

If ethnocentrism is associated with group conflict, the tendency to seek social rank higher than that of others—hierarchizing—is linked to individual conflict.

Human beings seek social rank, a fact that does not seem to be in dispute. In hunter-gatherer societies, "leveling" prevents the formation of obvious hierarchies—an individual who seeks dominance ("alpha animal" status) is automatically opposed by a coalition that deliberately denies him or her deference.[16,17,18] Social rank in these societies is therefore subtle and a product of personal qualities. For this reason, and because there can be little accumulation of material goods among peoples who move frequently and carry their belongings with them, hunter-gatherers generally do not have hereditary rank and wealth (with the exception of those who exploit a stationary, recurring resource, such as the Pacific Northwest Coast peoples of North America, whose harvesting of salmon permitted the development of complex societies without agriculture). However, once we look at societies that domesticate plants and animals we quickly see social hierarchy, often featuring massive social inequality. The implication is that although the tendency to hierarchize is indeed part of our evolutionary psychology, until relatively recently in our history this tendency did not create marked social inequality, social stratification, or despotism.[19] Hierarchizing does not necessarily generate hierarchy.

As individuals, our preoccupation with our relative standing seems to lead to a constant concern with managing the impressions others have of us, both minute-by-minute and in terms of our general reputation in our group.[20,21,22] Some of us are more skilled in reputation management than are others. Note, however, that social standing among human beings (and among chimpanzees and bonobos) is not primarily a matter of aggression or physical dominance, but involves political skill in coalition and support-building. It may appear that not everyone can be at the top of a hierarchy, so that our efforts to raise our own standing must necessarily lower that of others. However, our complex societies have numerous sets of criteria for the evaluation of relative standing. We can and often do choose a set of criteria in terms of which we personally can rank ourselves high, disregarding the fact that others may be using different criteria and in fact are ranking us as lower than themselves.[23,24] (Friendship often seems to have a component of hiding one's self-assessed "higher" rank from the other.*) We thus have numerous ways in which to build self-esteem.

Social scientists have tended to consider as economic many conflicts that, in fact, are about relative standing. From an evolutionary perspective, wealth is important in only two situations: when it is taken as an indicator of relative standing, and when we are dealing with such extreme poverty that differences in wealth affect whether one can care for oneself and one's family. Labor relations, for example, often have to do with symbolic markers of esteem and respect,

---

* One of our most highly regarded social skills, at least in Western societies, is the ability to convince others that we are (a) of high social standing, but that (b) we nevertheless regard the person we are talking to as our social equal (or nearly so). This ability is an important component of the skill-set often referred to as "charm."

and not simply salaries or job security. Even the economist's notion of physical capital has been broadened by sociologists, who now recognize the importance to relative social standing of both "cultural capital" (the kind of knowledge children of university professors, for example, acquire from their parents) and "social capital" (the network of cooperative relationships that can confer competitive advantage).[25,26] This expanded conception of capital is much more in line with the evolutionist's emphasis on relative standing and reputation than is the older, narrower notion of capital as money or property and their equivalents.

Though hierarchizing behavior certainly can lead to recurring problems such as bullying and damage to self-esteem, would we really wish to end all human interpersonal competition? After all, as many analysts have concluded, competition and conflict frequently lead to cooperation. Thus, a goal of minimizing the damage hierarchizing can do and maximizing its benefits for self-esteem and cooperation would appear to be more sensible than attempting to delete it entirely. Fortunately, the world has not been waiting for evolutionists to optimize the effects of conflict. There is, for example, the insight-filled academic specialization of conflict resolution (e.g., Deutsch and Coleman 2000). The findings and techniques of this very empirical field, as one would expect, appear to be quite compatible with what evolutionists now understand of our evolved human nature.

Lack of Match with Present Environment

To what extent do problems of survivability and sustainability reflect a mismatch between the environments to which evolution has adapted us and our present environments? Is modern society like an old-fashioned zoo, and we the animals locked in a world that was never made for us? The attractiveness of this argument is equaled only by its simplicity.

Laland, Odling-Smee and Feldman[27] have emphasized that we and other species construct much of our environment. Given that we everywhere have largely identical, evolution-produced brains, and given that physical reality imposes constraints on the possible, these constructed environments are doubtless much more similar to those of our ancestors than we might imagine. After all, today's human cultures and societies have numerous universal characteristics, a fact many of us have learned to ignore because of the powerful bias in the ethnographic literature towards "exoticizing" the other by focusing on what largely Western ethnographers have found unfamiliar. Crawford, in his valuable discussion of mismatch theory and ancestral environments, goes so far as to argue for *ancestralization*, defined as the "tendency for us to return to ancestral ways of behaving" once the ecological pressures that had caused a society to move "away from its ancestral form" have relaxed.[28] It is possible that there are ways in which our lived experience of post-industrial society is closer to that of our distant hunting-gathering forbearers than to that of our more recent agricultural ancestors. For example: on the one hand, the scale of our urban environments is vastly greater than that of any hunter-gatherer society; on the other, we seem to filter out unfamiliar places and people as we go about our daily lives, reducing our experienced environments to a more Pleistocene scope.

Nevertheless, there are numerous ways in which a mismatch between current and past environments may indeed be creating unhappiness and even pathology, both at the individual and societal level. John Bowlby, for example, arguably founded mismatch theory when he reasoned that the hospitals and nurseries of his time were far from our "environment of evolutionary adaptedness" and were interfering with the bonding of mother and infant, possibly creating subsequent pathologies of attachment behavior.[29] But mismatch theory has been chiefly applied in the fields of nutrition and medicine. There is at least some evidence, and considerable argument, that industrial food and grass-fed cattle are at least in part responsible for many of the ills of our time (including obesity, cardiovascular disease, and type 2 diabetes), and that the healthy diet is a "Paleo[lithic] diet".[30,31,32] Mismatch theory is a core component of the young field of Darwinian medicine, which seeks to understand the relationship between disease and evolutionary biology. The study of psychopathology, in particular, is in the process of being rethought from an evolutionary perspective.

Mismatch theory seems particularly pertinent when we compare the dangers and risks we face with those experienced by our Pleistocene ancestors. Our bodies and brains are well-adapted to cope with, for example, Pleistocene fauna. Our famous "fight-or-flee" physiology interrupts extraneous cognition and prepares our bodies to face physical danger. Threatened by a saber-toothed tiger (smilodon), it would presumably have been adaptive to flee, freeze, or possibly to fight. A scream of terror would have alerted other members of our group to danger, either bringing aid or at least permitting them (including relatives carrying copies of our genes) to escape. But note that these risks are all short-term.

Many of the risks of modern society are both long-term and evolutionarily novel. There are the health risks of alcohol and tobacco and unsafe sex and an industrial food diet, and the dangers of travel by automobile and airplane, for example. Today's risks are often ill-defined and collective rather than personal, as with problems of global warming, loss of biodiversity, genetic modification of organisms, bioterrorism, economic down-turns, the spread of nuclear weapons, and so forth.

The readiness with which we can dismiss quite real but long-term risks (e.g., risks of overeating, smoking, etc.) strongly suggests a mismatch between our Pleistocene brains and current reality. Let us imagine for a moment that our ancestors had evolved only to cope with long-term dangers and somehow had never faced predators or physical violence. Then suddenly, for the first time in our evolutionary history, there are lions among us! We would not have already evolved specialized adaptive mechanisms to cope with imminent violence because this threat would be evolutionarily novel and evolution is about past wisdom, so how would we react? Why, we would form committees to discuss courses of action, create lobbies and interest groups favoring some solutions but not others, and after a couple of decades we would still be shaking our heads over how many people were being devoured on our streets and blaming politicians for having failed to act.

This is precisely the situation we are in with respect to today's risks. Global warming is affecting us massively, we are destroying much of our planet's biodiversity, tens of millions die of HIV/AIDS. But the evolved mechanisms invoked are those that deal efficiently only with issues of coalitional politics and reputation, the kinds of longer-term problems that our Pleistocene ancestors *would* have faced. Only when a vague collective worry becomes a short-term personal threat to ourselves or those close to us do we react with the force these problems merit. We become members of a concerned and activist minority, the ethnocentrism reaction kicks in, and we enjoy a sense of moral superiority because we are now an in-group engaging in collective action, and we lament the moral inferiority of those who do not realize that their house is on fire. But the moral stance, as this paper began by arguing, is not enough.

Work-Arounds and a Praxis for Evolutionists

Praxis is the Greek term for acting or action, but it has come to mean theoretically, philosophically, theologically, or politically informed action. In the present context, "praxis" is intended to mean an *evolutionarily informed practice*. The idea requires the concept of "work-around," which originated with evolutionists Peter Richerson and Robert Boyd. They write that "The work- around hypothesis asserts that social instincts are part building-blocks and part constraints on the evolution of complex social systems."[33] Using the term "instincts" rather than "evolved psychology," they explain how work-arounds achieve social cooperation in large and complex units. Thus, the German Army during World War II was a highly effective fighting force because of an astute implicit understanding of human evolved psychology on the part of the generals. Divisions were organized on a territorial basis so that the men fought alongside other men who came from their own region and spoke with their own dialect and accent. Training emphasized bonding and loyalty between soldiers, and officers were enjoined to look after the welfare of their men to the fullest extent possible. In effect, the German Army consisted of units that at an emotional level were the kind of small and homogeneous units of kin in which our species likely evolved.[34] The "work around" consisted of a social organization in which a psychology evolved to defend one's own small band was channeled so it resulted in the support of the army of a highly aggressive modern state.

Experientially/intuitively-derived work-arounds of various scales are ubiquitous. Capitalism itself is in effect a grand system of work-arounds because capitalism rests on the channeling of social competition in the direction of resource production, acquisition, and control, leading to hyperproduction of goods and services. To simplify a more complex argument, when ambitious people compete for material wealth, capitalism flourishes and economies grow and far more is produced than is actually needed. When competition is channeled into, for example, religious knowledge and fervor as an alternative prestige strategy to wealth, then it is religion that grows, at least in part at the expense of economic development.[35] The work-arounds of the marketing/advertising industry are expert and endless in their manipulation of our evolved psychology. Evolved goals involving status and sexual competition and the nurturing of our families are worked into motivating us to purchase various goods and services. A concrete example: manufacturers of industrial foods lead us to purchase their products by loading them salt, sugar, and fat, taste preferences that would have been reliable guides to scarce and valuable nutrients in the Pleistocene. Today, of course, they lead us overnutrition. Another example: mass media are filled with the kinds of information that would have been adaptively relevant in small band societies, information about disasters and other dangers, the sexual activities of potential mates and rivals, the lack of reliability and honesty of some of those around us, and even the weather, among other topics. As with industrial food, an evolved taste is exploited in the pursuit of profit.

Politicians and corporations can do without explicit knowledge of our evolved psychology and the theory of work-arounds because they benefit from a great accumulation of empirical trial-and-error-derived knowledge. Their criteria for success, moreover, are unambiguous: power in the one case, profits in the other. Groups and individuals working to solve our collective problems of survivability and sustainability have neither of these advantages. They—we—therefore can and should make explicit of use our growing knowledge of evolutionary psychology to design deliberate work-arounds in our collective long-term interest.

For example, we tend to be nepotistic; that is, more altruistic towards real and symbolic kin than towards non-kin. A work-around that enlarged our perception of kin would expand the scope of our altruism. At the least, we should be able to make conservation as sexy as waste and mediation as admired as aggression; and we can at least teach our children to recognize and laugh at the would-be leader's standard speech, and to forgive themselves when they find it almost issuing from their own mouth.

Our knowledge of human evolutionary psychology is rapidly growing. As it grows, so too does the potential for an evolutionary praxis involving the deliberate design of work-arounds that meet our goals of sustainability and survivability. At the same time, mismatch theory gives us a valuable tool for understanding why we so often act collectively against our long-term interests, and why we as individuals so often act against our own future happiness and well-being. Meanwhile, evolutionary analyses of ethnocentrism and hierarchizing provide insight into both interpersonal and intergroup conflict. Yes, we do need to take a moral stance, and yes, we do need to study history; but we need to do both these things in the light of our evolved psychology, our human nature. Looking at ourselves with the evolutionist's gaze amounts, after all, to no more than restoring our species to the natural world.

Of course, this brief article has been more suggestive than substantive—it cannot substitute for a full-blown survey of evolutionary psychology and behavioral ecology and related fields. It has per force omitted major areas in which an evolutionary perspective can contribute to problem-solving, areas that include crime, the law, and governance. Similarly, although the discussion did touch on the perception of risk literature, it did not deal with the biases built into human thought by evolution, and the important roles of heuristics and bounded rationality in decision-making. We need to understand the rules of thumb we use in decision-making if we are to build effective work-arounds in the service of survival. Because reference was frequently made to the need for understanding our moral judgments concerning problems of survivability and sustainability, the evolutionary literature on the basis of human morality is certainly relevant. Because human decisions are in part emotional in nature, an evolutionary understanding of emotion would appear to be essential. Perhaps what is really needed is a reader, *Evolutionary Psychology for Policy-Makers.*

"Biology is not destiny unless we ignore it".[36] If we take an exclusively moral stance, then we make biology into destiny. If we take the perspective of an evolutionary praxis and use this understanding in crafting our policies, perhaps we can construct an enduring, sustainable way of life. Of course, as each new generation seeks to climb social hierarchies by demonstrating the superiority of their ideas and ways and plans over those of their elders, we will find that old solutions no longer seem to work and new ones are necessary. An evolutionary praxis requires, if not a permanent revolution, then at least a recurring one. Each generation must construct its own work-arounds, working around this tendency our species has to threaten its own survival.

Notes

1 Dennett, D. C. 1995. *Darwin's Dangerous Idea: Evolution and the Meanings of Life.* New York: Simon & Schuster.

2 Price, T. D. & A. B. Gebauer, eds. 1995. *Last Hunters—First Farmers: New Perspectives on the Prehistoric Transition to Agriculture.* Santa Fe, NM: School of American Research Press.

3 Tudge, C. 1999. *Neanderthals, Bandits and Farmers: How Agriculture Really Began.* New Haven, CT: Yale University Press.

4 Laland, K. N., J. Odling-Smee, & M. W. Feldman. 2001. Cultural niche construction and human evolution. *Journal of Evolutionary Biology* 14:22–33.

5 Simpson, G. G. 1965. *Tempo and Mode in Evolution.* New York: Hafner Pub. Co.

6 Rose, M. R. 1991. *Evolutionary Biology of Aging.* New York: Oxford University Press.

7 Barkow, J. H. 2006. Introduction: Sometimes the bus does stop. In: *Missing the Revolution: Darwinism for Social Scientists.* J.H. Barkow, ed. New York: Oxford University Press. In press.

8 Bunyard, P., Ed. 1996. *Gaia in Action: Science of the Living Earth.* Edinburgh, UK: Floris Books.

9 Lovelock, J. E. 1987. *Gaia: A New Look at Life on Earth,* New Edition. Oxford, UK: Oxford University Press.

10 Barkow, J. H. 1989. *Darwin, Sex, and Status: Biological Approaches to Mind and Culture.* Toronto: University of Toronto Press.

11 Barkow, J. H. 2000. Do extraterrestrials have sex (and intelligence)? In: *Evolutionary Perspectives on Human Reproductive Behavior,* vol. 907. D. LeCroy & P. Moller, eds. New York: Annals of the New York Academy of Sciences.

12 Irwin, C. J. 1987. A study of the evolution of ethnocentrism. In: *The Sociobiology of Ethnocentrism: Evolutionary Dimensions of Xenophobia, Discrimination, Racism and Nationalism.* V. Reynolds, V. Falger, & I. Vine, eds. London and Sydney: Croom Helm.

13 Gluckman, M. 1955. *Custom and Conflict in Africa.* Glencoe, Ill.:Free Press.

14 Kurzban, R. & M. R. Leary. 2001. Evolutionary origins of stigmatization: The functions of social exclusion. *Psychological Bulletin* 127:187–208.

15 Kurzban, R., J. Tooby, & L. Cosmides. 2001. Can race be erased? Coalitional computation and social categorization. *Proceedings of the National Academy of Sciences of the United States of America* 98:15387–15392.

16 Boaz, N. T. 2002. *Evolving Health: The Origins of Illness and How the Modern World is Making Us Sick.* New York: John Wiley & Sons.

17 Boehm, C. 1993. Egalitarian behavior and reverse dominance hierarchy. *Current Anthropology* 34:227–254.

18 Boehm, C. 2000. Conflict and the evolution of social control. In: *Evolutionary Origins of Morality: Cross-Disciplinary Perspectives.* L. D. Katz, Ed. Imprint Academic.

19 *Ibid.*

20 Barkow, J. H. 1989. *Darwin, Sex, and Status: Biological Approaches to Mind and Culture.* Toronto: University of Toronto Press.

21 Goffman, E. 1959. *Presentation of Self in Everyday Life.* Garden City, N.Y: Doubleday/Anchor.

22 Goffman, E. 1967. *Interaction Ritual: Essays in Face-to-Face Behavior.* Chicago: Aldine.

23 Barkow, J. H. 1975. Prestige and culture: A biosocial approach. *Current Anthropology* 16:553–572.

24 Barkow, J. H. 1989. *Darwin, Sex, and Status: Biological Approaches to Mind and Culture.* Toronto: University of Toronto Press.

25 Bourdieu, P. & J.C. Passeron. 1990. *Reproduction in education, society and culture,* 2nd edition. Theory, Culture and Society series. Thousand Oaks, CA: Sage Publications.

26 Fowler, B. 1997. *Pierre Bourdieu and Cultural Theory: Critical Investigations.* Thousand Oaks, CA: Sage Publications.

27 Laland, K. N., J. Odling-Smee, & M. W. Feldman. 2001. Cultural niche construction and human evolution. *Journal of Evolutionary Biology* 14:22–33.

28 Crawford, C. & D. L. Krebs, eds. 1998. *Handbook of Evolutionary Psychology: Ideas, Issues, and Applications.* Mahwah, New Jersey: Lawrence Erlbaum Associates.

29 Bowlby, J. A. 1969. *Attachment and Loss. Vol. I: Attachment.* New York: Basic Books.

30 Abrame, H. L. 1979. The relevance of paleolithic diet in determining contemporary nutritional needs. *Journal of Applied Nutrition* 31:43–59.

31 Burkitt, D. P. & S. B. Eaton. 1989. Putting the wrong fuel in the tank. Nutritional anthropology. *Contemporary Approaches to Diet and Culture* 5:189–191.

32 Cordain, L., B. A. Watkins, G. L. Florant, M. Kelher, L. Rogers, & Y. Li. 2002. Fatty acid analysis of wild ruminant tissues: evolutionary implications for reducing diet-related chronic disease. *European Journal of Clinical Nutrition* 56:181–191.

33 Richerson, P. J. & R. Boyd. 2001. Institutional evolution in the holocene: the rise of complex societies. In: *The Origin of Human Social Institutions.* W. G. Runciman, ed. Proceedings of the British Academy Vol. 110. Oxford, UK: Oxford University Press.

34 Barkow, J. H. 1989. *Darwin, Sex, and Status: Biological Approaches to Mind and Culture.* Toronto: University of Toronto Press.

35 Barkow, J. H. 1975. Prestige and culture: A biosocial approach. *Current Anthropology* 16:553–572.

36 Barkow, J. H. *Op. cit.*

Suggestions for Further Reading

Alcock, J. 2001. *The Triumph of Sociobiology.* New York: Oxford University Press.

Alexander, R. D. 1993. Biological considerations in the analysis of morality. In: *Evolutionary Ethics.* M. H. Nitecki & D. V. Nitecki, eds.: Albany, NY: SUNY Press.

Axelrod, R. 1984. *The Evolution of Cooperation*. New York: Basic Books.

Axelrod, R. & D. Dion. 1988. The further evolution of cooperation. *Science* 242:1385–1390.

Bailey, K. G. 2000. Evolution, kinship, and psychotherapy: Promoting psychological health through human relationships. In: *Genes on the Couch: Explorations in Evolutionary Psychotherapy*. P. Gilbert & K. G. Bailey, eds. New York, NY: Brunner-Routledge.

Bailey, K. G. & P. Gilbert. 2000. Evolutionary psychotherapy: Where to from here? In *Genes on the Couch: Explorations in Evolutionary Psychotherapy*. P. Gilbert & K. G. Bailey, eds. New York, NY: Brunner-Routledge.

Bailey, K. G. & H. E. Wood. 1998. Evolutionary kinship therapy: Basic principles and treatment implications. *British Journal of Medical Psychology* 71(4):509–523.

Barkow, J. H. 1975b. Strategies for self-esteem and prestige in Maradi, Niger Republic. In: *Psychological Anthropology*. T. R. Williams, ed. The Hague and Paris: Mouton.

–1989. *Darwin, Sex, and Status: Biological Approaches to Mind and Culture*. Toronto: University of Toronto Press.

–1992. Beneath new culture is old psychology. In: *The Adapted Mind. Evolutionary Psychology and the Generation of Culture*. J. H. Barkow, L. Cosmides & J. Tooby, eds. New York: Oxford University Press.

–2000. Our shared species-typical evolutionary psychology. Commentary on Laland, Odling-Smee and Feldman. *Behavioral and Brain Sciences* 23:148–149.

–2001. Universalien und evolutionäre psychologie. In: *Universalien und Konstruktivismus*. P. M. Hejl, ed. Frankfurt: Suhrkamp Verlag.

–2006. *Missing The Revolution: Darwinism For Social Scientists*. New York: Oxford University Press.

Barkow, J. H., L. Cosmides, & J. Tooby, eds. 1992. *The Adapted Mind: Evolutionary Psychology and the Generation of Culture*. New York: Oxford University Press.

Baron-Cohen, S. 1997. *The Maladapted Mind: Classic Readings in Evolutionary Psychopathology*. New York: Psychology Press.

Bateson, P. 1989. Does evolutionary biology contribute to ethics? *Biology & Philosophy* 4:287–302.

Beckstrom, J. H. 1993. *Darwinism Applied: Evolutionary Paths to Social Goals*. Westport, CT: Greenwood Publishing Group (Praeger).

–1997. Impact of the human egalitarian syndrome on Darwinian selection mechanics. *American Naturalist* 150 (supplement): 100–121.

Brown, D. E. 1991. *Human Universals*. New York: McGraw-Hill.

Browne, K. 1999. *Divided Labours: An Evolutionary View of Women at Work*. New Haven, CT: Yale University Press.

Browne, K. R. 2002. *Biology At Work: Rethinking Sexual Equality*. Piscataway, NJ: Rutgers University Press.

Burgess, R. L. & P. Draper. 1989. The explanation of family violence: The role of biological, behavioral, and cultural selection. In: *Family Violence.* L. Ohlin and M. Tonry, eds. Chicago and London: University of Chicago Press.

Buss, D. M. & N. M. Malamuth. 1996. *Sex, Power, Conflict: Evolutionary And Feminist Perspectives.* New York: Oxford University Press.

Buss, D. M. 1999. *Evolutionary Psychology: The New Science of the Mind.* Needham Heights, MA: Allyn & Bacon.

Campbell, D. T. 1965. Ethnocentric and other altruistic motives. In: *Nebraska Symposium on Motivation:* 1965. D. Levine, ed. Lincoln, Nebraska: University of Nebraska Press.

Charleton, B. 2000. *Psychiatry and the Human Condition.* Abingdon, England: Radcliffe Medical Press.

Charlton, B. G. 1997. A syllabus for evolutionary medicine. *Journal of the Royal Society of Medicine* 90:397–399.

Crawford, C. & D. L. Krebs, eds. 1998. *Handbook of Evolutionary Psychology: Ideas, Issues, and Applications.* Mahwah, New Jersey: Lawrence Erlbaum Associates.

Crawford, M. & D. Marsh. 1995. *Nutrition and Evolution.* New Canaan, CT: Keats Publishing.

Cronin, H. 1992. *The Ant and The Peacock: Altruism and Sexual Selection from Darwin to Today.* Cambridge and New York: Cambridge University Press.

Daly, M. & M. Wilson. 1998. *The Truth About Cinderella: A Darwinian View of Parental Love.* London: Weidenfeld & Nicholson.

Damasio, A. R. 1995. *Descartes' Error: Emotion, Reason, and the Human Brain.* New York: Avon Books.

Davis, H. & S. L. McLeod. 2003. Why humans value sensational news: An evolutionary perspective. *Evolution & Human Behavior,* 24, 208–216.

Dawkins, R. 1996. *River Out of Eden: A Darwinian View of Life.* New York: HCP / Basic Books.

Day, R. L., K. N. Laland, & F. J. Odling-Smee. 2003. Rethinking adaptation: the niche-construction perspective. *Perspectives in Biology and Medicine* 46(1), 80–95.

De Waal, F. 1996. *Good Natured: The Origins of Right and Wrong in Humans and Other Animals.* Cambridge, MA: Harvard University Press.

Deutsch, M. & P. T. Coleman, eds. 2000. *The Handbook of Conflict Resolution: Theory and Practice.* San Francisco, CA: Jossey-Bass.

Dimasio, A. 2000. *The Feeling of What Happens: Body and Emotion in the Making of Consciousness.* San Diego, CA: Harcourt/Harvest Books.

Eaton, S. B. 1992. Humans, lipids, and evolution. *Lipids* 27:814–820.

Eaton, S. B., M. C. Pike, R. V. Short, N. C. Lee, J. Trussell, R. A. Hatcher, J. W. Wood, C. M. Worthman, N. G. Blurton-Jones, M. J. Konner, K. R. Hill, R. Bailey, & A. M. Hurtado. 1994. Women's reproductive cancers in evolutionary context. *Quarterly Review of Biology* 69:353–367.

Eaton, S. B., M. Shostak, & M. Konner. 1988. *The Paleolithic Prescription: A Program of Diet & Exercise and A Design for Living.* New York: Harper & Row.

Ellis, L. & A. Walsh. 2000. *Criminology: A Global Perspective:* New York: Allyn & Bacon.

Fabrega, H. J. 1997. *Evolution of Sickness and Healing.* Berkeley/Los Angeles/London: University of California Press.

Fessler, D. M. T. 2002. The evolution of human emotions. In: *Oxford Encyclopedia of Evolution.* M. Pagel, ed. New York: Oxford University Press.

Flynn, J., P. Slovic, & H. Kunreuther, eds. 2001. *Risk, Media and Stigma: Understanding Public Challenges to Modern Science and Technology.* Sterling, VA: Stylus Publishing, LLC.

Frank, R. H. 1985. *Choosing the Right Pond: Human Behavior and the Quest for Status.* New York: Oxford University Press.

Gigerenzer, G. 2000. *Adaptive Thinking: Rationality In The Real World (Evolution and Cognition).* Oxford: Oxford University Press.

Gigerenzer, G. & R. Selten, eds. 2001. *Bounded Rationality: The Adaptive Toolbox.* Cambridge, MA: MIT Press.

Gilbert, P. & K. G. Bailey, eds. 2000. *Genes on the Couch: Explorations in Evolutionary Psychology.* New York: Brunner-Routledge.

Gil-White, F. J. 2001. Are ethnic groups biological "species" to the human brain? Essentialism in our cognition of some social categories. *Current Anthropology* 42, 4: 515–554.

Haig, D. 1993. Genetic conflicts in human pregnancy. *Quarterly Review of Biology* 68:495–532.

Harcourt, A. H. & F. M. B. de Waal. 1992. *Coalitions and Alliances in Humans and Other Animals.* Oxford: Oxford University Press.

Hirschfeld, L. 1996. *Race in the Making: Cognition, Culture, and the Child's Construction of Human Kinds.* Cambridge, MA: MIT Press.

Jaeger, C. C., O. Renn, E. A. Rosa, & T. Webler. 2001. *Risk, Uncertainty and Rational Action: Risk, Society and Policy.* Sterling, VA: Earthscan, Stylus Publishing, LLC.

Kahneman, D., A. Tversky, P. Slovic, & A. Tversky, eds. 1982. *Judgment under uncertainty.* Cambridge: Cambridge University Press.

Kalma, A. 1991. Hierarchisation and dominance assessment at first glance. *European Journal of Social Psychology* 21:165.

Katz, L. D., ed. 2000. *Evolutionary Origins of Morality: Cross-Disciplinary Perspectives.* Exeter, UK: Imprint Academic.

Lappé, M. 1994. *Evolutionary Medicine: Rethinking the Origins of Disease.* San Francisco: Sierra Club Books.

LeVine, R. A. & D. T. Campbell. 1972. *Ethnocentrism: Theories of Conflict, Ethnic Attitudes and Group Behavior.* New York: John Wiley and Sons.

Lockard, J. S. 1980. Speculations on the adaptive significance of self-deception. In: *The Evolution of Human Social Behavior.* J. S. Lockard, ed. New York: Elsevier.

Low, B. S. 1993. An evolutionary perspective on war. In: *Behavior, Culture, and Conflict in World Politics.* W. Zimmerman and H. K. Jacobson, eds. Ann Arbor: University of Michigan Press.

Masters, R. D. 1990. Evolutionary biology and political theory. *American Political Science Review* 84:195.

McGuire, M. T. 1992. Moralistic aggression, processing mechanisms, and the brain: biological foundations of the sense of justice. In: *The Sense of Justice: Biological Foundations of Law.* R. D. Masters & M. Gruter, Eds.. Newbury Park: Sage Publications.

McGuire, M. T. & A. Troisi. 1998. *Darwinian Psychiatry.* New York: Oxford University Press.

Mealey, L. 1985. The relationship between social status and biological success: A case study of the Mormon religious hierarchy. *Ethology and Sociobiology* 6:249–258.

Miller, B. D. 1993. *Sex and Gender Hierarchies.* Cambridge: Cambridge University Press.

Miller, G. 2000. *The Mating Mind: How Sexual Choice Shaped Human Nature.* New York: Doubleday.

Moore, A. J. 1993. Towards an evolutionary view of social dominance. *Animal Behaviour* 46:594–596.

Nesse, R., and G. Williams. 1994. *Why We Get Sick: The New Science of Darwinian Medicine.* New York: Times Books/Random House.

Nesse, R. M. 1989. Evolutionary explanations of emotions. *Human Nature* 1:261–289.

–2000. Is depression an adaptation. *Archives of General Psychiatry* 57:14–20.

Nesse, R. M. & G. C. Williams. 1998. Evolution and the origins of disease. *Scientific American* 279:86–93.

Nisbett, R. & L. Ross. 1980. *Human Inference: Strategies and Shortcomings of Social Judgment.* Englewood Cliffs, NJ: Prentice-Hall.

Nitecki, M. H., & D. V. Nitecki, eds. 1993. *Evolutionary Ethics.* Albany, NY: State University of New York Press.

Pani, L. 2000. Is there an evolutionary mismatch between the normal physiology of the human dopaminergic system and current environmental conditions in industrialized countries? *Molecular Psychiatry* 5:467–475.

Pinker, S. 2002. *The Blank Slate: The Modern Denial of Human Nature.* New York: Viking.

Reynolds, V., V. Falger, & I. Vine, eds. 1987. *The Sociobiology of Ethnocentrism. Evolutionary Dimensions of Xenophobia, Discrimination, Racism and Nationalism.* London and Sydney: Croom Helm.

Richerson, P. J. & R. Boyd. 1999. Complex societies: The evolutionary origins of a crude superorganism. *Human Nature* 10:253–289.

–2004. *Not by Genes Alone: How Culture Transformed Human Evolution.* Chicago: University of Chicago Press.

Ridley, M. 1997. *The Origins of Virtue: Human Instincts and the Evolution of Cooperation.* New York: Viking.

Rieber, R. W. 1991. *The Psychology of War and Peace: The Image of the Enemy.* New York: Plenum Press.

Salzman, P. C. 1999. Is inequality universal? *Current Anthropology* 40:31–61.

Segerstråle, U. 2000. *Defenders of the Truth: The Battle for Science in the Sociobiology Debate and Beyond.* New York: Oxford University Press.

Shaw, R. P. & Y. Wong. 1988. *Genetic Seeds of Warfare: Evolution, Nationalism, and Patriotism.* Winchester, Maine: Unwin Hyman.

Slovic, P. 1999. Trust, emotion, sex, politics, and science: surveying the risk-assessment battlefield. *Risk Analysis* 19:689–701.

–2000. *The Perception of Risk. Risk, Society and Policy.* Sterling, VA: Earthscan (Stylus Publishing LLC).

Spiro, M. E. 1996. Narcissus. *Ethos* 24:165–191.

Stearns, S. 1998. *Evolution in Health and Disease.* Oxford: Oxford University Press.

Stevens, A. & J. Price. 1996. *Evolutionary Psychiatry: A New Beginning.* London and New York: Routeledge.

Thienpont, K. & R. Cliquet. 1999. *In-group/Out-group Behaviour in Modern Societies: An Evolutionary Perspective.* Brussels: Vlaamse Gemeenschap.

Thornhill, R. & C. T. Palmer. 2000. *Rape: Biological Bases of Sexual Coercion.* Cambridge, MA: MIT Press.

Tooby, J. & L. Cosmides. 1990. The past explains the present: emotional adaptations and the structure of ancestral environments. *Ethology and Sociobiology* 11:375–424.

Trevathan, W. R., E. O. Smith, & J. J. McKenna. 1999. *Evolutionary Medicine.* New York: Oxford University Press.

Van Den Berghe, P. 1986. Skin color preference, sexual dimorphism and sexual selection: a case for gene-culture co-evolution? *Ethnic And Racial Studies* 9:87–113.

Vandermassen, G. 2005. *Who's Afraid of Charles Darwin? Debating Feminism and Evolutionary Theory.* Lanham, MD: Rowman & Littlefield Publishers.

Wallman, J. 1994. Vision—nature and nurture of myopia. *Nature* 371:201–202.

Walsh, A. 2006. Evolutionary psychology and criminal behavior. In: *Missing the Revolution: Darwinism for Social Scientists.* J.H. Barkow, ed. New York: Oxford University Press. In press.

Wiessner, P.& W. Schiefenhövel, eds. 1996. *Food and the Status Quest: An Interdisciplinary Perspective.* Providence, RI: Berghahn Books.

Williams, G. C., and R. M. Nesse. 1991. The Dawn of Darwinian medicine. *Quarterly Review of Biology* 66:1–22.

Wrangham, R.& D. Peterson. 1996. *Demonic Males: Apes and the Origins of Human Violence.* New York: Houghton Mifflin Company.

6

Belief and Survival

James E. Alcock

*Fear death?—to feel the fog in my throat, The
mist in my face, When the snows begin, and
the blasts denote I am nearing the place ...*

—Robert Browning, *Prospice*

Alpha and omega. Birth and death. Growth and decay. *Homo sapiens,* alone amongst species, is able to recognize this rhythm of life and the inevitability of personal demise. No matter how healthy or wealthy, or famous, or important one may be at the moment, we are all going to die. Yet, most people do not behave as though they live in the shadow of death, at least not until they encounter the metaphorical fog and mist. They want to live. They want to survive. And they take it so much for granted that everyone else should also want to live—with the possible exception of those trapped by severe disability or racked by intractable pain—that they view suicidal thought or action as indicative of mental illness.

However, in much of the world, human mortality is all too evident, and too often the result of war or famine or disease, rather than old age. Yet, in modern Western societies, we have become disconnected from death, and we work hard to avoid confronting it. We delegate the task of dealing with the dying and the deceased to hospitals and funeral homes. Many people, having satisfied more basic needs for food and shelter, turn their energies to the sometimes narcissistic pursuit of health and fitness, striving to stave off any signs that they are "nearing the place." Plastic surgery, silicone implants, botox injections, hair transplants, liposuction—modern medicine offers a cornucopia of techniques to help keep us looking young and to help us avoid dealing with mortal reality. And where modern medicine disappoints, both ancient and very new "alternative" therapies offer succour in the attempt to preserve youth and health.

However, die we ultimately must. And to deal with the existential angst that this realization produces, societies everywhere have constructed belief systems—religions, cults, philosophies—to assuage anxiety about the prospect of personal demise, and to soften the pain felt at another's death. Many of these belief systems deal with death essentially through denial, teaching that we do not die, but instead "pass away" to some form of eternal life.

And so it has been throughout human history, this overriding concern for personal survival. However, until recent times, we have never had to think very much about threats to the survival of our species, or about the sustainability of the physical environment upon which we depend for our existence. Yet, largely because of products of the astounding scientific and technological progress of the past century, and the major societal changes that they have helped produce, we now face a plethora of planetary perils: threats posed by overpopulation in some regions and precipitously falling birth rates in others, deleterious environmental changes, resource depletion, pollution, nuclear and biological weapons, and new forms of pestilence.

Given the ubiquitous human concern for personal survival, one might expect the citizens of the world to be just as concerned about their collective survival, and to demand that their leaders work together to confront these perils. Such is not the case. Concerns about collective survival fall for the most part on deaf ears. Why should this be so? No doubt a number of factors account for the disconnect between concerns about personal and collective survival, but a key element is *belief.*

What Is a Belief?

A belief, simply put, is an expectancy about something: "If I step off this diving board, I will fall into the water." "If I eat that peach, I will have an enjoyable gustatory sensation." "If we don't do something about global warming, we will face catastrophic consequences." Our beliefs serve a very important function: based on our own past experience and what we have learned from others, they guide us in interpreting and reacting to the world around us. It would be an impossible task to treat each situation that confronts us as though it were entirely novel. We would have no time for productive thought if each time we held something in our hands, we had to consider whether or not it might fall were we to let go of it; or if each time we came to a door, we had to figure out what it was, decide how to go about opening it, and wonder about what would happen when we did. We react quickly and appropriately to our surroundings, for the most part, because we carry with us a set of beliefs about the nature of those surroundings, and about the probable effects of each particular action that we might take.

Beliefs not only guide our behaviours, but also motivate them: "I believe that I have already studied enough to do well in the examination," leads to a decision to take the rest of the evening off. "I believe that this is going to be a horrible examination, and I have to know everything perfectly," is more likely to result in intense study. "No matter how much I study, I won't be able to pass the examination," may lead to despair, and even to giving up on studying altogether. Similarly, a belief that recycling helps spare the environment may promote recycling behaviour, while a belief that recycling is expensive and does not reduce our garbage problems may do just the opposite.

Each of us undoubtedly harbours some beliefs that are simply untrue. The problem is that they do not seem any different in quality from others that are true, and so we cannot root them out even if we wanted to do so. So it is that other's beliefs seem factual and true to them, no matter how wrong-headed they may seem to us. In terms of its effect on behaviour, it is not particularly important whether or not a belief is correct, for faulty or irrational beliefs motivate particular behaviours just as surely as do those that are correct and rational. Ironically, while erroneous or irrational beliefs generally produce ineffective or counterproductive behaviours, such behaviours may lead to outcomes that nonetheless reinforce the faulty belief. The student who believes he or she is not going to be able to pass the examination may not devote the necessary time to studying, and the self-fulfilling prophecy plays itself through when the student fails. If enough people believe that recycling is a waste of time, the recycling campaign ultimately fails, thus giving apparent verification to the prior belief. Much of modern clinical psychological treatment involves focussing on an individual's "irrational" beliefs, and helping him or her to challenge them and to construct a system of more rational beliefs that promotes productive attitudes and behaviour.

Some beliefs also serve the important function of reducing anxiety. The worry that follows a serious injury is reduced, at least for a time, if we believe that modern medicine will cure our ills. The anxiety associated with death is decreased by belief in an immortal soul. The fears raised by warnings about the dangers of greenhouse gases, global warming, and acid rain are reduced if we believe that these warnings are gross exaggerations, or that, if the problems are real, scientists will solve them for us. Such mitigating beliefs are based partly in hope or blind faith and partly in denial. Those based in hope may be very adaptive at times, for they fend off feelings of helplessness and depression and may even allow the channelling of one's energies into finding effective ways to deal with the danger. (Think of the high morale of the British during the Second World War, based partly on a belief that, somehow, "Britons never ever ever shall be slaves."*) A belief based in denial may also reduce anxiety in the short term, but it is more likely to be maladaptive in the long run, because instead of motivating action to overcome the threat, it calms people into doing nothing. If the threat is not real, why should one do anything? (Consider Chamberlain's dealings with Hitler in Berlin.) If we can persuade ourselves that the disappearing ozone layer is not really a threat, then surely we do not need to worry or do anything about it.

Acquisition of Beliefs

Beliefs are obviously of great importance, but they are certainly not all rooted in fact, even though we generally treat them as though they are. What, then, determines what we come to believe? How is it that some come to believe that humanity is facing very serious threats to its survival, while others totally ignore or even gainsay such a belief? How is it that some believe that they can be successful in changing society, while others take the view that no matter what they do, their actions will not change anything?

Although we humans like to think that we can reason most things out, we are not born to logic; we have to acquire it. Neither are we born with the concept of truth; that is a human invention, and we have to learn it, too. The newborn baby can distinguish very little in its sensory surroundings, although its brain is set up to be more sensitive and responsive to some stimuli than to others. What it is born

---

* From the de facto British national anthem, *Rule Britannia.*

with is a remarkable neurological apparatus that is capable of processing the huge amounts of information pouring in from the world outside, finding patterns in this sensory stream, and then using those patterns to guide behaviour and aid in the processing of further information. Some of the apparent patterns are meaningless; most are not. Gradually, associations develop amongst objects and events, and these associations ultimately give rise to beliefs. The learning of these associations, which is referred to as *experiential learning,* is based primarily on two factors: stimulus similarity and temporal contiguity.

Stimulus Similarity

What we learn about one stimulus and its relationships with other stimuli and events is automatically generalized to stimuli that resemble it. The baby who is soothed and nourished by an adult will associate that positive outcome with other similar adults, at least until old enough to be able to discriminate particular adults (mother, father) from adults in general. The child who is frightened by a particular dog will initially be frightened of all dogs and of other animals that resemble dogs. The infant begins a process that continues throughout life, the development of an extensive set of categories, or "schemas," built on similarities amongst classes of objects or events. As a result, we can react to novel stimuli on the basis of the category that most closely corresponds to it. If we see something with a tail and fins swimming in the water, we can quickly categorize it as fish or shark. This categorization will not only influence our further perception and subsequent memory (making the object more clearly fish-like or shark-like, even if we did not see it so clearly), but it will also determine our reaction, especially if we happen to be in the water at the time! By being able to categorize a novel stimulus in terms of an existing schema, we can then react to it on the basis of our system of beliefs about the schema—for example, "Sharks are dangerous, get out of the water."

Temporal Contiguity

Our nervous systems are set up to treat two events that occur closely together in time ("temporally contiguous") as though the first is a predictor of the second. Touch a hot stove, feel pain—the lesson is a rapid one, and we immediately learn to associate pain with touching the stove. Kick a soda pop machine that has not delivered even though we have paid, and when the soda pop is suddenly expelled, we "learn" that the kick dislodged the drink, and the next time we encounter a similar problem, we are likely to respond in the same way. Of course, as we grow, we learn to respond differentially depending on the presence or absence of other stimuli. The older child can distinguish between a stove that is likely to be hot and one that is not.

Although acquiring an association between two events happens readily, especially if one of the events is aversive, the association is not so readily undone. There is a fundamental asymmetry in the nervous system with regard to the role of temporal contiguity, one that is very adaptive in terms of individual survival even if it is sometimes counterproductive with regard to truth.[1,2] If a child is burned by a stove on one occasion, and then subsequently touches the stove by accident and is not burned, the association between stove and pain is not reset to zero by that second event. Obviously, this is a good thing in the main, for avoiding harm is vital to our survival, and it would not generally serve us well if we quickly unlearned such associations. However, it also provides the basis for magical thinking. Magical thinking occurs when we assume that there is a causal link between two events, but have no knowledge of or interest in the nature of the causal link. Wear a lucky charm, and win the tennis match. Take vitamin C, and the cold symptoms decrease. Such thinking persists primarily for one reason—the asymmetry in our nervous systems mentioned earlier. When there is a "hit"—we wear the charm and win the match, or we take the pill and the cold improves—we remember that. When there is no hit, we downplay it and forget it. Such "partial reinforcement" sets up an even more powerful association, one that is more resistant to being broken down by non-reinforcement in future. The proclivity for magical thinking is part and parcel of our makeup whether we like it or not. It is tied to the way our brain makes associations between events.

Experiential learning is based on short-term outcomes. We learn quickly to avoid actions that lead to pain, and to repeat actions that lead to pleasure. However, problems associated with issues of collective survival generally involve long-term consequences to individual action, which means that no experiential learning is going to occur. This is also true with regard to some issues of personal health: A good example of this is the ways in which smokers put their health at risk, despite wanting to survive and stay healthy. An association is learned between the act of smoking and a pleasant feeling. In terms of direct experience, then, the brain learns that smoking a cigarette is followed by a positive outcome. (Ultimately, because of physiological addiction, the brain also learns that not smoking is associated with an unpleasant outcome.) If cancer showed up on the lips or the lungs of a smoker shortly after smoking, it is doubtful that anyone would smoke. The delayed consequences turn whatever association there is between smoking and cancer into an intellectual one, something that emerges from epidemiology and statistics, and not from experience.

Likewise, despite the obvious human penchant for individual survival, many people abandon medical treatments that are almost certain to save them from very undesirable long-term consequences. Consider glaucoma, a progressive eye disease that is the second leading cause of blindness in North America. In its early stages, it is all but undetectable by the person who has it, but it can be controlled if detected and treated early, preserving precious vision. The typical treatment involves putting drops in one's eyes every day for the rest of one's life. Yet non-compliance with a treatment regimen is reported to be a leading cause of glaucoma blindness.[3,4] Many people stop taking the medication and deny that there is a problem because neither the risks of *not* taking the treatment nor the benefits of taking it are apparent in the short run.

Similar considerations apply to most of the threats to our collective survival. Wasting energy, polluting the air, overusing resources—these activities do not produce immediate and obvious negative consequences. The link between action and outcome is one that can only be apprehended by the intellectual process, and not by the experiential.

Of course, we also process information on an *intellectual* level. While our experience teaches us that touching a hot stove will produce pain, our intellectual learning teaches us about the concepts of heat and insulation, so that it makes sense to us that we should not feel pain if we touch the stove while wearing an oven mitt. In every society, adults devote considerable energy to both the formal and informal teaching of beliefs about objects and events to their children. We also acquire a great deal of information through *social learning*—by watching what happens to others. We can learn how to go through a revolving door by watching someone else go before us. We can learn that certain classes of people are to be feared by being aware of our parents' anxiety in their presence. We can learn about what is likely to accrue to us as a result of particular actions by watching what happens to others who act in that particular way.

Experiential and intellectual processing generally work in a complementary manner. However, this is not always so. We may read about the reduction in the ozone layer, and yet at an experiential/emotional level find it hard to believe that the dear old sun above can do bad things to us. Despite the warnings about ultraviolet rays, tanning salons seem to flourish and beaches are filled with sun worshippers.

Belief and Authority

While beliefs are often based either in direct experience or are products of reason, as discussed above, we also acquire many of our beliefs, including some of our most important ones, directly from authority. We may like to think that we are reasonable and logical, and that we can always defend our beliefs with logic or data, or both. However, that simply is not the case most of the time. For example, virtually everyone believes that the Earth is a globe; whereas a few centuries ago virtually everyone believed that it was flat. We may smile at the benighted views of our ancestors, but can any of us truly defend our belief that the world is a globe, without relying ultimately on some authority or another? The Earth certainly appears to be flat, at least in many places in the world. We may cite as evidence the old story of watching the tall ship as it disappears over the horizon, but it is doubtful if any of us have personally had that experience. We may rely on photographs taken from space, but the pictures themselves could easily have been contrived. Ultimately, we take those pictures to be real because we rely on the authority of the astronauts who go into space and of the news organizations who tell us about them. Our daily lives are filled with such authorities: teachers, commentators, newspapers, books, "experts."

How, without reliance on authorities, would the average person ever know that the ozone layer is shrinking, or that some fish stocks are in perilous decline, or that greenhouse gases will lead to deleterious effects on weather? Worse, many modern issues are very complex, and we may often be overwhelmed with too much information, some of it conflicting, leaving us unable to understand what we have been told or what we should do. Even highly respected authorities may present us with quite disparate views, and so whom do we believe? One scientist tells us that the ozone layer is rapidly eroding, and that global warming will have dire consequences. Another scientist disputes this conclusion, while a politician decries the Kyoto Accord on greenhouse emissions. How are we to choose our experts, given that such matters are all but impossible for us to sort out on our own? The best we can usually do is to choose which experts to believe on the basis of their credentials and perceived bias, or lack thereof.

Resistance to Change

If we believe that the world faces a sustainability crisis—that the future of the human species is threatened—how can we present information to others who do not share this belief in such a way that they will give it serious consideration? If the information produces anxiety, people may react with denial, or with blind faith that their leaders and scientists will deal with the problem. Indeed, there are a number of factors that produce resistance to messages that challenge our current beliefs, even if the message is well-supported by good evidence:

1. *Maintaining consistency:* In general, at least in Western societies, children are taught implicitly and explicitly to avoid self-contradiction; as a result, consistency is generally what we prefer. What happens when we hold two beliefs that are mutually inconsistent? Festinger built his Theory of Cognitive Dissonance around this question. He argued, and there has been a long history of research that has for the most part supported his view, that when we have two beliefs, or "cognitions," that are inconsistent, we experience discomfort ("dissonance"); this motivates us to take steps to bring these dissonant cognitions into consistency, especially if they are important to us.[5] For example, the smoker who believes that smoking causes cancer has two cognitions that are dissonant: "Smoking causes cancer" and "I smoke." To reduce the dissonance, this person may do such things as: (a) stop smoking; (b) devalue the message or the authority of its source; (c) deliberately seek out information that throws the link with cancer into doubt while avoiding information that underscores the danger; (d) promise him/herself to stop next week or next year; or (e) rationalize that a cure for cancer is just around the corner. Similarly, the owner of a gas-guzzling sports utility vehicle may experience dissonance if we persuade him or her that operating

such a vehicle is detrimental to the environment. If we are successful in our persuasion, the resulting dissonance may cause the person to consider disposing of the vehicle, but that may well be an unattractive option. The individual might be more likely to reduce dissonance either by downplaying our message, or choosing amongst options similar to those that pertained to the smoker above.

2. *Primitive beliefs:* Primitive beliefs are beliefs so fundamental to our belief system that we take them as axiomatic.[6] They may also be so widely held in our social group that no one else challenges them either (e.g., the importance of democracy and freedom), or they may be subjective and cannot be assessed by others (e.g., "I feel God's presence in my life"). In any case, they serve as screening instruments for new information, and inconsistent information may be ignored or downplayed. We accept without question that objects cannot remain suspended in the air if there is nothing holding them up: if one were accidentally to let go of a coffee cup, and instead of falling, it stayed suspended in mid-air, no one would be likely to say, "That's interesting—so, sometimes objects don't fall to the ground." Not only would the individual be shocked, but if he or she later reported this information to others, it would not pass muster with their primitive belief about gravitation. Those hearing such a claim would be more likely to view the report as a sign of a mental disorder, a hallucination. Similarly, thirty years ago, a fisher who had grown up believing that the oceans offer an inexhaustible supply of fish may very well have disregarded or viewed with suspicion warnings that the seas were being over-fished.

3. *Emotion and belief:* Some beliefs are associated with strong emotional feelings, and such beliefs are typically very resistant to change. Smokers who view their habit as an expression of individual freedom may react very emotionally to attempts to persuade them that their second-hand smoke may be harmful to others, viewing this as an attack on their freedom.

4. *Interconnectedness of beliefs:* Our beliefs, for the most part, do not stand alone from each other; they are interconnected. One's belief in the gravitational force is very much connected to one's belief about what happens when one hang glides or parachutes or dives. As a consequence, changes in one belief will usually involve changes in interconnected beliefs, and this may make a

belief resistant to change. This is especially so with important beliefs, and with beliefs that are more central (that is, having a very high number of connections to other important beliefs). A devout Roman Catholic who reasons that it is not logical for the Church to oppose the use of condoms when the world is threatened by AIDS will, by virtue of that reasoning, come to the conclusion that the Pope, who opposes the use of condoms, is in error. However, this would mean the doctrine of Papal infallibility must also be called into question. Yet, were the Pope not infallible, then many other important doctrinal beliefs based in Papal authority would also be thrown into question. It may be too overwhelming for some people to perform such a housecleaning of their belief system, and therefore easier not to tamper with the belief about condoms. Similarly, if we come to believe that by driving our gas-guzzling vehicle, we are harming the environment, then to be consistent, we will likely have to modify a number of other related beliefs (for example, we may have to accept that our motor boat pollutes the waters). This may require an undesirable or unacceptable level of behaviour change, and so instead we choose to defend the gas-guzzling vehicle and to deny the environmental threat.

5. *Dogmatism:* Dogmatism is a personality feature that greatly influences how resistant an individual will be to information that opposes his or her beliefs.[7] An "open-minded" person can take in, evaluate and act upon new information, without being overly influenced by irrelevant internal factors (habits, primitive beliefs) or external factors (such as social pressure). An individual who is more closed-minded, or "dogmatic," is less open to new information that conflicts with primitive beliefs, and more influenced by the consideration of the source of the information, rather than the actual information itself.

7. *Groupthink:* Of course, just as birds of a feather flock together, people usually prefer the company of those whose belief systems are not widely discrepant with their own. This minimizes social friction and cognitive dissonance, but as a result, it also minimizes the opportunity to examine new and contrary information that might lead to reassessment of one's beliefs. Indeed, in group decision-making situations, this can even lead to " groupthink," where members of a highly cohesive group of decision-makers strive to reach a consensus, even at the cost of ignoring

inconsistent information and failing to consider all alternative courses of action.[8] The groupthink phenomenon has been used to explain how various disastrous historical events came about, such as the Challenger space shuttle explosion, the ill-conceived Bay of Pigs invasion, and the escalation of the U.S. involvement in Vietnam, where it is likely that none of the participants would have come to the same decision individually.

Beliefs and Collective Survival

When we confront global threats to our collective survival, we must change people's beliefs if we are going to motivate them to take effective action. Of course, in the short run, changing their beliefs is likely to increase anxiety, leading to the resistance discussed above.

There are at least three generic beliefs that need to be in place before individuals are likely to take any action:

1. *A belief that there is a problem.* People must know about the problem and accept that the problem is a genuine threat. It is one thing to state that the ozone layer is shrinking, but it is quite another for members of the public to understand what that means. Why is the ozone layer important? Few people have the technical skills or knowledge to evaluate such a statement or the data that might support or contradict it. We learn from experience—our own and especially that of others—that when a tornado or blizzard is headed our way, we are in some peril and need to take appropriate action. We do not have any history of experience with the kinds of global threats we now face.

   Once again, the individual must rely on authority, and one must weigh the pronouncements of various scientists, politicians, environmentalists, and others, and then arrive at some belief about how serious the threat is. That is why, in the battle to persuade people, it is vital to build up authoritative sources of information that can be trusted not to be manipulative or unnecessarily alarmist. It is also important to avoid inducing too much fear, for information campaigns based on inducing fear usually do not work very well. In order to reduce anxiety, people either avoid a fear-inducing message or reinterpret it in a less threatening way—for example, as "extremist propaganda."

2. *A belief that a solution exists.* Even if people come to recognize a threat, before their behaviours will change they have to believe that the situation is reversible and that one has the means to do

something about it. If a tornado is coming, get out of its way. If war erupts, join the army, buy victory bonds, rally round the flag. But what about global warming? What about overpopulation? What about terrorism? Is there anything to be done? People must be informed not just about the threats, but about the potential solutions, or they will simply avoid the message in order to reduce anxiety.

3. *A belief that one has a role to play in solving the problem.* Even if we believe that there is a serious threat to our collective survival, and even if we believe that the problem is solvable, we may not believe that we have any personal role to play in solving it. Whether it is local street crime or international violence, whether it is an AIDS epidemic or extinction of fish stocks, individuals may feel that they are helpless to do anything about these problems.

People need to be persuaded that there is indeed something that they can do about these global threats to survival. Information campaigns must provide clear instructions about reasonable behaviours that will help respond to the threat. Here, the role of environmental and other groups is important, for such groups provide not only direction to individual members, but also create a sense of collective power. Moreover, such groups can establish themselves as reliable authorities regarding such issues, just as Médecins Sans Frontières, the International Red Cross, and Amnesty International have done in their respective areas of interest. As well, being part of a movement is much more reassuring and welcoming to most individuals than trying to change the world alone.

Individual Versus Collective Rationality

The problem of motivating individuals to action is complicated by a fundamental social problem, sometimes referred to as a "collective dilemma" or a "many-one" problem. Situations in which an individual's actions, be they socially desirable or not, have very little influence on the overall, collective outcome typically reinforce the individual for *not* contributing to the common good.[9] Indeed, in many important social circumstances, individual and collective rationality are actually opposed, at least in the short term. Consider the following mundane example: the number of cottages on a formerly pristine lake has grown to the point that the effluent being dumped into the lake is now noticeable and unpleasant and a health threat to the community of cottagers. Yet the effluent produced by any individual cottage is so small that, by itself, it would not be a problem, and its absence would not be a benefit. If the conscientious cottager, aware of the growing pollution of the lake, acts in a prosocial manner and installs a septic tank, this involves a real cost to him or her, even though the action itself may well have no discernable effect on the overall quality of the lake water. If no one else follows suit, the lake remains polluted, and this noble cottager is left with both a polluted lake and the cost of installing and servicing the septic tank. However, if a majority of cottagers make the environmentally appropriate decision and install septic tanks, then the quality of the lake water improves, and the fact that one cottager continues to empty sewage into the lake has only a negligible effect on the water quality. Thus, from a position of pure individual economic rationality, the best strategy is to do nothing. If the others do not act, the situation will not improve anyway, and one has saved some money. If the others do act, the lake gets cleaned up anyway, and one has again saved some money.

Collective dilemmas are difficult to resolve because individual (economic) rationality is pitted against what is best for the collectivity. This led Hardin, who studied the problem of overgrazing on English Commons in the 1500s, to conclude that individuals are unlikely to be persuaded to change their individual behaviours without the imposition of social controls such as criminal sanctions or taxes—"mutual coercion mutually agreed upon."[10] This is particularly pertinent to environmental issues. One motorist using leaded gas is unlikely to have a measurable negative impact on anything if most other motorists use unleaded gas, just as one car using unleaded gas is unlikely to have a measurable positive impact if most other motorists use leaded gas.

Obviously, there is more than pure economic consideration at play here. For some, the increased self-esteem in doing what is collectively the right thing will be profit enough, but history shows the folly of counting on this as a reliable motivator for most people. That is why, of course, governments sensitive to environmental and health issues have banned leaded gas outright.

Another complication on the global level is that what is desirable in developed countries may not be so in non-developed countries; the reinforcement contingencies are often quite different. As just one example, overpopulation is generally not a problem in modern industrialized countries because having fewer children is generally linked to a number of positive outcomes, including greater personal freedom and more family wealth. Yet, for a peasant in an underdeveloped region of the world, this is not likely to be the case. First of all, the peasant may have been taught that only a fraction of babies—say two out of five—survive childhood, and therefore, one needs to have many more births than one wants children. It takes a long time to observe the effects of improved health care and sanitation on infant mortality rates. Second, the peasant's long-term future depends on the ability to rear healthy sons who will be able to care for their parents in old age. If one gives any thought at all to birthrate, it may be, "Who will take care of me when I am old, if I don't have lots of children?" Thus, to such a peasant, arguments about the problems of overpopulation may have little appeal.

The Importance of Science

In face of the many serious problems that plague our species, it is troubling that for many people, the value of science and logic appears to have declined significantly over the past twenty-five years. For some, science is irrelevant; for others, it is seen as the cause of our problems in the first place. For yet others, magical belief systems have taken root that interfere with finding real solutions, even though they may reduce anxiety. At bottom, science remains a quest for knowledge, for truth, a quest armed with methodology painstakingly developed over the past few centuries that helps us to obtain knowledge and, more importantly, helps us to fend off self-delusion. This pursuit is vital to our survival.

In the battle for collective survival in the face of growing global threats, we cannot rely on science alone. We must understand and attempt to change the beliefs of our citizenry. For the reasons discussed above, this will not be an easy task. Providing accurate information by itself will not modify beliefs. However, by applying what is known about belief formation and belief change, there is a very good chance that people can be persuaded to change their beliefs and their behaviours, and work together for the common good, so that we can keep Browning's metaphorical fog and mist from enveloping our planet.

Notes

1 Alcock, J.E. 1995. The belief engine. *Skeptical Inquirer* 19:14–18.
2 Alcock, J.E. 1996. The propensity to believe. In: *The Flight From Reason.* N. Levitt, et al., eds. New York: New York Academy of Sciences.
3 Goldberg, I. 2000. Compliance with medical management in glaucoma. *The Asian Journal of Ophthalmology* 2 (4): 3–6.
4 Van Buskirk, E.M. 1986. The compliance factor [Editorial]. *American Journal of Ophthalmology* 10: 609–610.
5 Festinger, L. 1957. *A Theory of Cognitive Dissonance.* Stanford, CA: Stanford University Press.
6 Rokeach, M. 1960. *The Open and Closed Mind.* New York: Basic Books.
7 Ibid.
8 Janis, I. 1972. *Victims of Groupthink.* Boston, MA: Houghton Mifflin.
9 Alcock, J.E. & D. Mansell. 1977. Predisposition and behaviour in a collective dilemma. *Journal of Conflict Resolution* 21: 443 457.
10 Hardin, G. 1968. The tragedy of the commons. *Science* 162: 1243–1248.

7

The Social Psychology of Sustainability

David G. Myers

Can we move nations and people in the direction of sustainability? Such a move would be a modification of society comparable in scale to only two other changes: the Agricultural Revolution and the Industrial Revolution of the past two centuries. Those revolutions were gradual, spontaneous, and largely unconscious. This one will have to be a fully conscious operation. ... If we actually do it, the undertaking will be absolutely unique in humanity's stay on the Earth.

—William D. Ruckelshaus
Former Environmental Protection
Agency director
"Toward a Sustainable World," 1989

Life is good. Today we enjoy luxuries unknown even to royalty in centuries past—hot showers, flush toilets, microwave ovens, jet travel, big screen television, e-mail, and Post-it notes. But on the horizon beyond the sunny skies, dark clouds are forming. In scientific gatherings hosted by the United Nations, the Royal Society (UK), and the National Academy of Sciences, a consensus has emerged: increasing population and increasing consumption have combined to overshoot the earth's ecological carrying capacity.[1,2] We are spending our environmental capital, not just living off the interest.

In 1950, the earth carried 2.5 billion people and 50 million cars. Today it has more than 6 billion people and ten times as many cars. If world economic growth enabled all countries to match Americans' present car ownership, the number of cars would multiply yet another thirteen times over.[3] These cars, along with the burning of coal and oil to generate electricity and heat homes, produce greenhouse gases that contribute to global warming. The warmest 23 years on record have all occurred since 1975, with 1998 the hottest year ever (likely surpassed by the time you read this). The polar icecaps are melting at an accelerating rate; so are mountain glaciers from the Alps to the Andes. Africa's Mount Kilimanjaro has lost 82 percent of its 1912 icecap; at the present melting rate it will disappear by 2016.[4] Equatorial insects and vegetation are migrating toward the poles. With the changing climate, extreme atmospheric events—heat waves, droughts, and floods—are becoming more common. As precipitation falls more as rain and less as snow, the likely result will be more floods in rainy seasons and less melting snow and glaciers for rivers during dry seasons.

With world economic growth and population both destined to increase (even as birth rates fall), resource depletion and global warming now seem inevitable. Ergo, the need for more sustainable consumption has taken on "urgency and global significance".[5] The simple, stubborn fact is that the earth cannot indefinitely support our present consumption, much less the expected increase in consumption. For our species to survive and flourish, some things must change.

If we are to cease stealing from our descendants, we can first increase technological and agricultural efficiency. As cool fluorescent light bulbs consume less than hot incandescent bulbs and e-mail consumes less than printed and delivered letters, so we can look forward to dry, heatless ultrasound washing machines, to cars driven by fuel cells that produce water exhaust, and to roofs and roads that double as solar energy collectors.

But we must, second, also modify our human behavior. It is we humans who drive gas-slurping SUVs, eat grain-slurping cattle, and operate tree-slurping deforestation machines. And it is we humans who, according to the *Wall Street Journal,* are failing to respond to eco-marketing. "After a decade of designing products to appeal to environmentally friendly sensibilities, many companies have concluded that 'green' sales pitches don't sell. As they head to stores, Americans are leaving their consciences at the landfills." Plastic soda bottle recycling has dropped by two-thirds since 1995, while single-serve bottle sales doubled to 18 billion in 2000.[6]

Having overshot the earth's carrying capacity, and with our material appetites still swelling—as people in developing countries desire to join us in having nicer clothes, cars, and computers, and as Westerners seek ever more CDs, air conditioning, and vacation condos—how can we change direction?

Population control will help. As food security improves and more women become educated and empowered, birth rates fall. Alas, even if the birth rates of every country fell to replacement levels, the bulge of younger humans would feed more population growth for years to come. So we also need to moderate consumption. But how?

One avenue is policies that provide incentives for conservation. What we tax we get less of. What we reward we get more of. So, if our roads are clogged and our air polluted, we can create fast highway lanes that reward car pooling. We can build bike lanes and subsidize mass transit. We can follow the European example of shifting some of our taxes toward fuel.

In *Luxury Fever,* economist Robert Frank proposes we go a step further and create a progressive consumption tax that rewards saving and investment, and increases the price of nonessential luxury goods. Tax people, he suggests, not on what they earn, but on what they spend—their earnings minus their savings and perhaps their charity. If tax rates were low for subsistence spending, but rose to 70 percent for consumption over $500,000 a year, people who would have bought an 80-foot yacht might now get by with a 50-foot sailboat. Those considering a 3000-square-foot home might settle comfortably into a cozier 2000-square-foot home.[7]

Support for such policies will require a shift in public consciousness not unlike that occurring during the 1960s civil rights movement and the 1970s women's movement. As the atmosphere warms and oil and other resources become scarce, such a shift is inevitable. Is there any hope that, before the crisis becomes acute, human priorities might shift from accumulating money to finding meaning, and from aggressive consumption to nurturing connections? Perhaps social psychology can help by exposing our materialism, by informing people of the disconnect between economic growth and human morale, and by helping people understand why materialism and money fail to satisfy.

Increasing Materialism

Does money buy happiness? No! Ah, but would a *little* more money make us a *little* happier? Many of us smirk and nod. There is, we believe, *some* connection between fiscal fitness and feeling fantastic. Most folks queried by Gallup report that, yes, they would like to be rich. Three in four entering American collegians—nearly double the 1970 proportion—now consider it "very important" or "essential" that they become "very well off financially" (Figure 1). It is not just collegians. Asked by Roper pollsters to identify what makes "the good life," 38 percent of Americans in 1975, and 63 percent in 1996, chose "a lot of money."[8] Money matters.

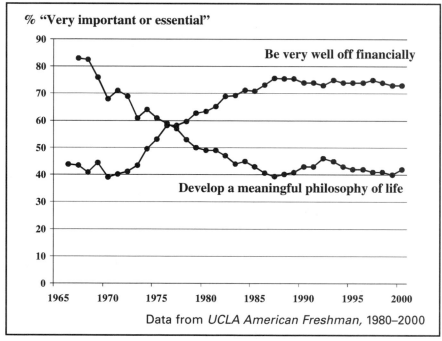

% "Very important or essential"

Be very well off financially

Develop a meaningful philosophy of life

Data from *UCLA American Freshman, 1980–2000*

Figure 1. The changing materialism of entering collegians. *UCLA American Freshman* annual surveys of more than 200,000 students revealed an increasing desire for wealth between 1970 and the mid 1980s, with continued high materialism thereafter.

It's the old American dream: life, liberty, and the purchase of happiness. "Of course money buys happiness," wrote Andrew Tobias. Wouldn't anyone be happier with the indulgences promised by the magazine sweepstakes: a 40-foot yacht, deluxe motor home, private housekeeper? Anyone who has seen *Lifestyles of the Rich and*

Famous or the Travel Channel's *Ultimate Luxuries* knows as much. "Whoever said money can't buy happiness isn't spending it right," proclaimed a Lexus ad. No wonder many people hunger to know the secrets of "the millionaire mind."

Ironically, however, those who most ardently seek after money tend to live with lower well-being, a finding that "comes through very strongly in every culture I've looked at," reports research psychologist Richard Ryan.[9] Ryan's collaborator, Tim Kasser, concludes from their studies that people who instead seek "intimacy, personal growth, and contribution to the community" experience greater quality of life.[10] Their research, summarized in Kasser's *The High Price of Materialism,* echoes an earlier finding that college alumni with "Yuppie values"—those preferring high income and vocational success to close friendships and marriage—were twice as likely as their former classmates to be "fairly" or "very" unhappy.[11, 12]

Consider: what was the most satisfying event in your last month? When Ken Sheldon and his colleagues put such questions to university students, and asked them also to rate how much ten different needs were met by the satisfying event, the students were most likely to report that the event met their needs for self-esteem, relatedness to others, and autonomy. At the bottom of the list of satisfaction predictors was money and luxury.[13]

Economic Growth and Human Morale

Materialism—lusting for more—exacts ecological and psychic costs. But might actually getting more pay emotional dividends? Perhaps unsustainable consumption is bad for the planet, but good for one's sense of well-being. Would people be happier if they could trade a simple lifestyle for one with a private chef, Aspen ski vacations, and travel on private jets? Would they be happier if they won a publishers' sweepstake and could decide among its suggested indulgences, including a 40-foot yacht, deluxe motor home, and luxury car?

We can observe the traffic between wealth and well-being by asking first if rich nations have more satisfied people. There is, indeed, some correlation between national wealth and well-being (indexed as self-reported happiness and life satisfaction). The Scandinavians have been mostly prosperous and satisfied; the Bulgarians are neither. But early 1990s data revealed that once nations reached about $10,000 GNP per person, which was roughly the economic level of Ireland, higher levels of national wealth were not predictive of increased well-being. Better to be Irish than Bulgarian. But whether one was an average Irish person or West German (with double the Irish purchasing power) hardly mattered.[14, 15]

Secondly, we can ask whether within any given nation, rich people are happier. In poor countries—where low income threatens basic needs—being relatively well off does predict greater well-being.[16] In affluent countries, where most can afford life's necessities, affluence matters less. Income increases and windfalls temporarily boost happiness, and recessions create short-term psychic losses. But over time the emotions wane. Once comfortable, more and more money produces diminishing long-term returns. World values researcher Ronald Inglehart therefore found the income-happiness correlation to be "surprisingly weak."[17] David Lykken illustrates: "People who go to work in their overalls and on the bus are just as happy, on the average, as those in suits who drive to work in their own Mercedes."[18]

Even the super rich—the *Forbes* 100 wealthiest Americans—report only slightly greater happiness than average.[19] And even winning a state lottery seems not to enduringly elevate well-being.[20] Such jolts of joy have "a short half-life," notes Richard Ryan.[21]

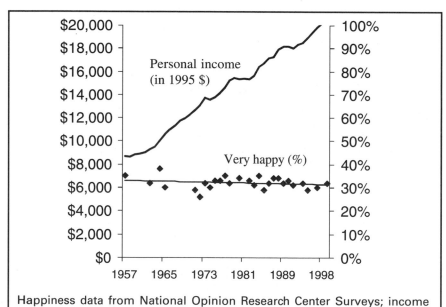

Happiness data from National Opinion Research Center Surveys; income data from *Historical Statistics of the United States* and economic indicators.

Figure 2. Money and happiness. Although average disposable income has more than doubled since the 1950s, the average American's reported happiness has remained almost unchanged.

Third, we can ask whether, over time, a culture's happiness rises with its affluence. Does our collective well-being float upward with a rising economic tide?

In 1957, as economist John Galbraith was describing the United States as *The Affluent Society,* Americans' per person income was (in 1995 dollars) about $9000. Today, as Figure 2 indicates, the United States is the doubly affluent society. Although this rising tide has lifted the yachts faster than the dinghies, nearly all boats have risen. With double the spending power, thanks partly to the surge in married women's employment, we now own twice as many cars per person, eat out twice as often, and are supported by a whole new world of technology. Since 1960, we have also seen the proportion of households with dishwashers rise from 7 to 50 percent, with clothes dryers rise from 20 to 71 percent, and with air conditioning rise from 15 to 73 percent. Daily showers—a luxury enjoyed by 29 percent of Americans in 1950—had become a necessity for 75 percent of Americans by 1999.[22]

So, believing that it's "very important" to "be very well-off financially," and having become better off financially, are today's Americans happier? Are we happier with espresso coffee, caller ID, and suitcases on wheels than we were before?

We are not. Since 1957, the number of Americans who say they are "very happy" has declined from 35 to 32 percent. Meanwhile, the divorce rate has doubled, the teen suicide rate has more than doubled, the violent crime rate has tripled (even after the recent decline), and more people than ever (especially teens and young adults) are depressed.

I call this soaring wealth and shrinking spirit "the American paradox." More than ever, we have big houses and broken homes, high incomes and low morale, secured rights and diminished civility. We excel at making a living but often fail at making a life. We celebrate our prosperity but yearn for purpose. We cherish our freedoms but long for connection. In an age of plenty, we feel spiritual hunger.

These facts of life explode a bombshell underneath our society's materialism: *Economic growth has provided no boost to human morale.*

Why Materialism and Money Fail to Satisfy

So why have we not become happier? Why have 41 percent of Americans—up from 13 percent in 1973—come to regard automotive air-conditioning as "a necessity"?[23] Why do yesterday's luxuries so quickly become today's requirements and tomorrow's relics? Two principles drive this psychology of consumption:

Our human capacity for adaptation. The "adaptation-level phenomenon" is our tendency to judge our experience (for example, of sounds, lights, or income) relative to a neutral level defined by our prior experience. We adjust our neutral levels—the points at which sounds seem neither loud nor soft, temperatures neither hot nor cold, events neither pleasant nor unpleasant—based on our experience. We then notice and react to up or down changes from these levels.

Adaptation researcher Allen Parducci recalls a striking example: "On the Micronesian island of Ponope, which is almost on the equator, I was told of a bitter night back in 1915 when the temperature dropped to a record-breaking 69 degrees [Fahrenheit]!"[24] In the United States, Midwesterners, perhaps after watching too many Baywatch episodes, see sunny California as a happy place to live. But contrary to Midwesterners' intuitions, Californians—much as they may prefer their climate—are no happier.[25] Warm sun is just your average day.

Thus, as our achievements rise above past levels, we feel successful and satisfied. As our social prestige, income, or in-home technology surges, we feel pleasure. Before long, however, we adapt. What once felt good comes to register as neutral, and what formerly was neutral now feels like deprivation. As the good feelings wane, it takes a higher high to rejuice the joy.

So, could we ever create a social paradise? Donald Campbell answered no: if you woke up tomorrow to your utopia—perhaps a world with no bills, no ills, someone who loves you unreservedly—you would feel euphoric, for a time. Yet before long, you would recalibrate your adaptation level and again sometimes feel gratified (when achievements surpass expectations), sometimes feel deprived (when they fall below), and sometimes feel neutral.[26] That helps explain why, despite the realities of triumph and tragedy, million-dollar lottery winners and people who are paralyzed report roughly similar levels of happiness. It also explains why material wants can be insatiable—why many a child "needs" just one more Nintendo game. Or why Imelda Marcos, surrounded by poverty while living in splendor as wife of the Philippines' president, bought 1060 pairs of shoes. When the victor belongs to the spoils and the possessor is possessed by possessions, adaptation level has run amuck.

Our wanting to compare. Much of life revolves around "social comparison." We are always comparing ourselves with others. And whether we feel good or bad depends on who those others are. We are slow-witted or clumsy only when others are smart or agile. Let one baseball player sign a new contract for $15 million a year and his $8 million teammate may now feel less satisfied. "Our poverty became a reality. Not because of our having less, but by our neighbors having more," recalled Will Campbell in *Brother to a Dragonfly*.

Social comparisons help us understand the modest income-happiness correlation. Middle- and upper-income people in a given country, who can compare themselves with the relatively poor, tend to be slightly more satisfied with life than their less fortunate compatriots. Nevertheless, once people reach a moderate income level, further increases do little to increase their happiness. Why? Because we tend to compare upward as we climb the ladder of success or income.[27, 28] Beggars compare with more successful beggars, not millionaires, noted Bertrand Russell in 1930's *The Conquest of Happiness*. Thus, "Napoleon envied Caesar, Caesar envied Alexander, and Alexander, I daresay, envied Hercules, who never existed. You cannot, therefore, get away from envy by means of success alone, for there will always be in history or legend some person even more successful than you are."[29]

Rising income inequality, notes Michael Hagerty, makes for more people markedly above us in our communities. And that helps explain why those living in communities with a large rich-poor gap tend to feel less satisfied.[30] If you live in a 2000 square foot house in a community filled with other 2000 square foot houses you likely are happier than if living in the same house amid 4000 square foot homes. Television modeling of the lifestyles of the wealthy also serves to accentuate feelings of "relative deprivation" and desires for more.[31] The writer of Ecclesiastes understood this: "I have also learned why people work so hard to succeed: It is because they envy the things their neighbors have. But it is useless. It is like chasing the wind. ... It is better to have only a little, with peace of mind, than be busy all the time with both hands, trying to catch the wind."

The adaptation level and social comparison phenomena give us pause. They imply that the quest for happiness through material achievement requires continually expanding affluence. But the good news is that adaptation to simpler lives can also happen. If choice or necessity shrinks our consumption we will initially feel pain, but it will pass. "Weeping may tarry for the night, but joy comes with the morning," reflected the Psalmist. Indeed, thanks to our capacity to adapt and to adjust comparisons, the emotional impact of significant life events—losing a job, or even a disabling accident—dissipates sooner than most people suppose.[32]

Toward Sustainability and Survival

A shift to post-materialist values will gain momentum as people

- face the implications of population and consumption growth for climate change, habitat destruction, and resource depletion;

- realize that materialist values mark *less* happy lives; and

- appreciate that economic growth has *not* bred increased well-being.

"If the world is to change for the better it must have a change in human consciousness," said Czech poet-president Vaclav Havel. We must discover "a deeper sense of responsibility toward the world, which means responsibility toward something higher than self."[33] If people came to believe that stacks of unplayed CDs, closets full of seldom-worn clothes, and garages with luxury cars do not define the good life, then might a shift in consciousness become possible? Instead of being an indicator of social status, might conspicuous consumption become gauche?

Social psychology's contribution to a sustainable and survivable future will come partly through its consciousness-transforming insights into adaptation and comparison. These insights also come from experiments that lower people's comparison standards and thereby cool luxury fever and renew contentment. In two such experiments, Marshall Dermer and his colleagues put university women through imaginative exercises in deprivation. After viewing depictions of the grimness of Milwaukee life in 1900, or after imagining and writing about being burned and disfigured, the women expressed greater satisfaction with their own lives.[34]

In another experiment, Jennifer Crocker and Lisa Gallo found that people who five times completed the sentence "I'm glad I'm not a ..." afterward felt less depressed and more satisfied with their lives than did those who completed sentences beginning "I wish I were a ... ."[35] Realizing that others have it worse helps us count our blessings. "I cried because I had no shoes," says a Persian proverb, "until I met a man who had no feet."

Social psychology also contributes to a sustainable and survivable future through its explorations of the good life. If materialism does not enhance life quality, what does? One ingredient of well-being is the satisfaction of our deep "need to belong." As social creatures, we are deeply motivated not only to eat, to procreate, and to achieve, but also to bond with important others. Our distant ancestors who formed attachments were more likely to reproduce and to stay together to nurture their offspring to maturity. In solo combat, our ancestors were not the toughest predators. But as hunters they learned that six hands were better than two. Those who foraged in groups also gained protection from predators and enemies. The inevitable result: an innately social creature. That helps us understand why those supported by intimate friendships or a committed marriage are much likelier to declare themselves happy. In National Opinion Research Center surveys of over 40,000 randomly sampled Americans since 1972, 40 percent of married adults, and 23 percent of never married adults, have declared themselves "very happy."

The same surveys revealed that 27 percent of those rarely or never attending religious services declared themselves very happy, as did 47 percent of those attending multiple times weekly. Positive traits—self-esteem, an internal locus of control, optimism, and extraversion—also mark happy times and lives.[36]

Leisure and work experiences that engage one's skills also mark happy lives. Between the anxiety of being stressed and overwhelmed and the apathy of being bored lies a zone in which people experience "flow," notes Mihaly Csikszentmihalyi. Flow is an optimal state of absorption during which we lose consciousness of time and self. When people's experience is sampled using electronic pagers, they report greatest enjoyment not when mindlessly passive, but when unselfconsciously absorbed in a challenge. What's more, the *less* expensive (and usually more involving) a leisure activity is, the happier people are while engaged in it. Most of us are happier talking to friends than watching TV. Low consumption recreations prove most satisfying.[37,38]

And that is indeed good news. The things that make for a genuinely good life—close, supportive relationships, a hope-filled faith community, positive traits, engaging activity—are enduringly sustainable.

Notes

1 Heap, B., & J Kent, eds. 2000. *Towards Sustainable Consumption: A European Perspective.* London: The Royal Society.

2 Oskamp, S. 2000. A sustainable future for humanity? How Can Psychology Help? *American Psychologist* 55: 496–508.

3 Myers, N. 2000. Sustainable consumption: The meta-problem. In: *Towards Sustainable Consumption: A European Perspective.* B. Heap & J. Kent, eds. London: The Royal Society.

4 Revkin, A. C. Glacier loss seen as clear sign of human role in global warming. *New York Times.* February 19, 2001. (accessed at nytimes.com).

5 Heap, B., & J Kent, eds. 2000. *Towards Sustainable Consumption: A European Perspective.* London: The Royal Society.

6 *Wall Street Journal.* 2002. Green sales pitch isn't moving any products, March 6: B1, B4.

7 Frank, R. 1999. *Luxury Fever: Why Money Fails to Satisfy In An Era of Excess.* New York: The Free Press.

8 Putnam, R. 2000. *Bowling Alone.* New York: Simon & Schuster.

9 Ryan, R. In pursuit of affluence, at a high price. *New York Times.* February 2, 1999 (via www.nytimes.com).

10 Kasser, T. 2000. Two versions of the American dream: which goals and values make for a high quality of life? In: *Advances in Quality of Life: Theory and Research.* E. Diener & D. Rahtz, eds. Dordrecth, Netherlands: Kluwer.

11 Kasser, T. 2002. *The High Price of Materialism.* Cambridge, MA: MIT Press.

12 Perkins, H. W. 1991. Religious commitment, yuppie values, and well-being in post-collegiate life. *Review of Religious Research* 32: 244–251.

13 Sheldon, K. M., A.J. Elliot, K. Youngmee, & T. Kasser. 2001. What is satisfying about satisfying events? Testing 10 candidate psychological needs. *Journal of Personality and Social Psychology* 80: 325–339.

14 Inglehart, R. 1990. *Culture Shift in Advanced Industrial Society.* Princeton, NJ: Princeton University Press.

15 Inglehart, R. 1997. *Modernization and Postmodernization: Cultural, Economic, and Political Change in Societies.* Princeton, NJ: Princeton University Press.

16 Argyle, M. 1999. Causes and correlates of happiness. In: *Foundations of Hedonic Psychology: Scientific Perspectives on Enjoyment and Suffering,* D. Kahneman, E. Diener, & N. Schwartz, eds. New York: Russell Sage Foundation.

17 Inglehart, R. 1990. *Culture Shift in Advanced Industrial Society.* Princeton, NJ: Princeton University Press.

18 Lykken, D. T. 1999. *Happiness.* New York: Golden Books.

19 Diener, E., J. Horwitz, & R.A.Emmons. 1985. Happiness of the very wealthy. *Social Indicators* 16: 263–274.

20 Brickman, P., D. Coates, & R.J. Janoff-Bulman. 1978. Lottery winners

and accident victims: is happiness relative? *Journal of Personality and Social Psychology* 36: 917–927.

21 Ryan, R. In pursuit of affluence, at a high price. *New York Times.* February 2, 1999. (via www.nytimes.com).

22 Myers, D. G. 2000. *The American Paradox: Spiritual Hunger In An Age of Plenty.* New Haven, CT: Yale University Press.

23 Schor, J. B. 1998. *The Overworked American.* New York: Basic Books.

24 Parducci, A. 1995. *Happiness, Pleasure, and Judgment: The Contextual Theory and Its Applications.* Hillsdale, NJ: Erlbaum.

25 Schkade, D.A., & D. Kahneman. 1998. Does living in California make people happy? A focusing illusion in judgments of life satisfaction. *Psychological Science* 9: 340–346.

26 Campbell, D.T. 1975. The conflict between social and biological evolution and the concept of original sin. *Zygon* 10: 234–249.

27 Gruder, C. L. 1977. Choice of comparison persons in evaluating oneself. In: *Social Comparison Processes.* J.M. Suls & R.L. Miller, eds. Washington, D.C.: Hemisphere Publishing.

28 Suls, J. & F. Tesch. 1978. Students' preferences for information about their test performance: a social comparison study. *Journal of Applied Social Psychology* 8: 189–197.

29 Russell, B. 1930. *The Conquest of Happiness.* London: Unwin Paperbacks.

30 Hagerty, M.R. 2000. Social comparisons of income in one's community: evidence from national surveys of income and happiness. *Journal of Personality and Social Psychology* 78: 764–771.

31 Schor, J. B. 1998. *The Overworked American.* New York: Basic Books.

32 Gilbert, D.T., E.C. Pinel, T.D. Wilson, S.J.Blumberg, & T.P. Wheatley. 1998. Immune neglect: a source of durability bias in affective forecasting. *Journal of Personality and Social Psychology* 75: 617–638.

33 Havel, V. 1990. *Disturbing the Peace.* New York: Knopf.

34 Dermer, M., S. J.Cohen, E. Jacobsen, & E. A. Anderson. 1979. Evaluative judgments of aspects of life as a function of vicarious exposure to hedonic extremes. *Journal of Personality and Social Psychology* 37: 247–260.

35 Crocker, J., & L. Gallo. 1985. *The self-enhancing effect of downward comparison.* Paper presented at the American Psychological Association convention.

36 Myers, D.G. 2000. The funds, friends, and faith of happy people. *American Psychologist* 55: 56–67.

37 Csikszentmihalyi, M. 1990. *Flow: The Psychology of Optimal Experience.* New York: Harper & Row.

38 Csikszentmihalyi, M. 1999. If we are so rich, why aren't we happy? *American Psychologist* 54: 821–827.

8

The Need for a Planetary Ethic*

Ervin Laszlo

Values and beliefs determine the way we perceive the world and suggest the ways we prioritize the responses to our perceptions. They are essential for the survival for our species; they affect almost all areas of human behavior. However, in the wider context of society, individual values and beliefs are often at odds, unlikely to conform to a common standard on their own. But attempting to impose values and beliefs "from above" would be misguided and, as historical experience testifies, likely to produce negative results. Self-regulation is just as important in the values sphere as it is in any other sphere of society. If we are to create better chances of human survival, we need to introduce some self-regulatory mechanism in civil society. In a democratic context this must be a *moral* mechanism—more exactly, a moral code or principle. And in a globally interdependent world, it must be a universally acceptable and spontaneously shared moral code or principle: a *planetary ethic*.

A planetary ethic is a major imperative of our time. We all have our private morality, our personal ethic. This varies with the personality, the ambitions, and the circumstances of each of us. It

* Adapted from Ervin Laszlo, *Macroshift: Navigating the Transformation to a Sustainable World.* San Francisco: Berret-Koehler. 2001.

reflects our unique background, heritage, and family and community situation. We also have a public morality, the ethic shared in our community, ethnic group, state, or nation. This is the ethic the group in which we live requires of us in order for it to function. It reflects its culture, social structure, economic development, and environmental conditions. But there is also a universal morality—a planetary ethic. This is the ethic the human family as a whole requires so that all its members can live and develop.

Universal morality is an essential part of private and public morality. It respects the conditions under which all people in the global community can live in dignity and freedom without destroying each other's chances of livelihood, culture, society, and environment. It does not prescribe the nature of our private and public morality—it only ensures that they do not give rise to behaviors that are damaging to the planetary community that is the vital context of our lives.

How could a morality shared the world over arise and spread in society? Traditionally, setting the norms of morality was the task of the religions. The Ten Commandments of Jews and Christians, the provisions for the faithful in Islam, and the Rules of Right Livelihood of the Buddhists are examples. Today the dominance of science has reduced the power of religious doctrines to regulate human behavior, and many people look to science for practical guidance. Yet scientists, with some notable exceptions, discover few principles that would provide a basis for universal morality. Saint-Simon in the late 1700s, Auguste Comte in the early 1800s, and Émile Durkheim in the late 1800s and early 1900s all tried to develop "positive" scientific observation- and experiment-based principles for a meaningful and publicly acceptable ethic. This endeavor as a whole, however, was so foreign to science's commitment to value neutrality and objectivity that it was not taken up by mainstream twentieth century scientists.

In the final decade of the twentieth century, scientists as well as political leaders began to recognize the need for principles that would suggest universal norms of behavior. In the April 1990 "Universal Declaration of Human Responsibilities," the InterAction Council, a group of twenty-four former heads of state or government, expressed this conviction: "Because global interdependence demands that we must live with each other in harmony, human beings need rules and constraints. Ethics are the minimum standards that make a collective life possible. Without ethics and self-restraint that are their result, humankind would revert to the survival of the fittest. The world is in need of an ethical base on which to stand."

The Union of Concerned Scientists, an organization of leading scientists and citizens, concurred. "A new ethic is required," claimed a statement signed in 1993 by 1670 scientists from seventy countries, including 102 Nobel laureates. "This ethic must motivate a great movement, convincing reluctant leaders and reluctant governments and reluctant peoples themselves to effect the needed changes." The scientists noted our new responsibility for caring for the Earth, and warned that, "a great change in our stewardship of the Earth and the life on it is required if vast human misery is to be avoided and our global home on this planet is not to be irretrievably mutilated." Human beings and the natural world, they said, are on a collision course. This may so alter the living world that it will be unable to sustain life as we know it.

In November of 2003 a group of Nobel laureates meeting in Rome stated, "Ethics in the relations between nations and in government policies is of paramount importance. Nations must treat other nations as they wish to be treated. The most powerful nations must remember that as they do, so shall others do." And in November of 2004 the same group of Nobel laureates declared, "Only by reaffirming our shared ethical values—respect for human rights and fundamental freedoms—and by observing democratic principles, within and amongst countries, can terrorism be defeated. We must address the root causes of terrorism—poverty, ignorance and injustice—rather than responding to violence with violence."

Undoubtedly the time has come to give serious attention to a morality that can be embraced by all people regardless of their creed, religion, race, sex, or secular belief. It must have intuitive appeal, addressing the basic moral instinct present in all healthy individuals. This merits serious thought. Because the egalitarian ideals of Marx, Lenin, and Mao failed in the practice of communist countries, the highest expression of everyday ethics for the great bulk of humanity has been liberalism, the conceptual heritage of Bentham, Locke, Hume, and the classical school of British philosophers. Here ethics and morality have no objective basis: human actions are based on self-interest, moderated at best by an element of altruistic sympathy. People are not to be prevented from pursuing their self-interest as long as they observe the rules that permit life in civilized society. "Live and let live" is the liberal principle. You can live in any way you please, as long as you do not break the law.

In today's world, classical liberalism makes for a misplaced form of tolerance. Letting everyone live as they please as long as they keep within the law entails a serious risk. The rich and the powerful could consume a disproportionate share of the resources to which the poor, too, have a legitimate claim, and both rich and poor could inflict irreversible damage on the environment that we have to share with them.

Rather than "live and let live," we need a planetary ethic that is just as intuitively meaningful and instinctively appealing as the ethic of liberalism, but better adapted to current conditions on this planet. Such an ethic would substitute for liberalism's "live and let live" Gandhi's exhortation to "live more simply, so others can simply live." This idea needs further refinement, however, because we are not concerned with the intrinsic simplicity of lifestyles but with their impact on society and nature. This must not exceed the capacity of the planet to provide for the needs of all its inhabitants. In consequence, the planetary ethic we need is better stated as: "live in a way that allows others to live as well."

Living in ways that enable others to live as well is the planetary ethic of our time, but is it practicable? Will it be accepted and embraced by a significant segment of society? This question will not be decided by moral philosophers, but by processes within democratic societies. The times when kings, popes, and princes could decide what is moral and what is not are over. In a democratic world, principles regulating people's behavior come from the people themselves.

Thomas Jefferson said that if you believe the people are not sufficiently informed to exercise the power of demos in society, the democratic solution is not to *take* power from their hands, but to *inform* them. Informing others of the requirement for an ethic adapted to our time is not a quixotic endeavor. If people realize that there is a real need for a planetary ethic, and that abiding by it does not dictate the nature of our private and public morality or entail undue sacrifice, they will respond with interest and alacrity.

The need for a planetary ethic is real, and its relevance to human survival can be made evident. Human life is intimately tied to the lives of other species; in fact, to the entire biosphere. If we continue to interfere with the ecological balance established among the diverse species, conditions in the biosphere will evolve along pathways distinctly inhospitable to humankind's well-being and threatening for its survival. Agricultural lands will erode, weather patterns will turn hostile, water tables will fall and ocean levels rise, lethal radiation will penetrate the atmosphere, and micro-organisms fundamentally incompatible with our organism will proliferate. A wide variety of ecocatastrophes will come about.

We can also make clear that abiding by a planetary ethic does not entail particular sacrifices. Living in a way that enables others in the biosphere to live as well does not mean being self-denying; we can continue to strive for excellence and beauty, personal growth and enjoyment, even for comfort and luxury. But in the context of a planetary ethic, the pleasures and achievements of life are defined in relation to the quality of enjoyment and level of satisfaction they provide rather than in terms of the amount of money they cost and the quantity of materials and energy they require. This ethic requires that we take into account the basic question, "Is how I live and what I do compatible with the right to life of others?" Does it allow access to the basic resources of life for six and a half billion humans, and for the plants and animals that populate our life-supporting environment?

These questions must be answered by each of us in regard to everything we do. They can be answered using a basic rule of thumb: envisage the consequences of your action on the life and activity of others. Does it or does it not rob basic resources from them? Does it or does it not despoil their environment? These questions are not impossible to answer. By way of example, we should look at three of the most widespread practices in the contemporary world: eating meat, smoking, and the use of the private automobile.

Eating Meat

Cutting back on our consumption of meat is both a sustainability and a health imperative. World meat consumption has risen from 44 million tons in 1950 to 217 million tons by 1999, nearly a fivefold increase—an untenable trend. In addition, the meat we buy today is not the safe meat grandmother bought in 1950. Quite aside from the danger of it being infected by mad cow disease, it is likely to contain progesterone, testosterone, avoparcin, and clenbuterol—chemicals farmers pump into cattle to fatten them up and keep them healthy. Anabolic steroids, growth hormones, and beta-agonists turn fat into muscle; antibiotics stimulate growth and protect sedentary animals against diseases they would not get if they were kept in more natural conditions.

A diet based on heavy meat eating is not only unhealthy, it is immoral: it indulges a personal fancy at the expense of depleting resources essential to feed the entire human population. Red meat comes from cattle, and cattle must be fed. The grain fed to cattle is removed from human consumption. If cows returned equivalent nutrition in the form of meat, their feed would not be wasted. But the calorific energy provided by beef is only one-seventh of the energy of the feed. This means that in the process of converting grain into beef, cows "waste" six-sevenths of the nutritional value of the planet's primary produce. The proportion is more favorable in poultry: an average chicken uses only two-thirds of the calorific value of the feed it consumes.

There is simply not enough grain to feed all the animals that would be needed to supply meat for the tables of the world's entire population. These giant herds of cattle and endless farms of poultry would require more grain than the total output of the agricultural lands—according to some calculations, about twice as much. Given the amount of land available for farming and the known and presently used agricultural methods, doubling today's grain production would call for economically prohibitive investments. The rational and moral solution is to phase out the mass-production of cattle and poultry—not by massive slaughter but by breeding fewer animals and breeding them healthier.

The nutritive needs of the entire human population can be satisfied by eating more vegetables and grain and less meat, using first and foremost the produce of one's own country, region, and environment. Grain- and plant-based food self-reliance provides a healthier diet, and it allows the world's economically exploitable agricultural lands to be worked to satisfy the needs of the whole of the human family.

Smoking

What goes for meat eating also goes for smoking. The fact that smoking is dangerous to health can be read on every packet of cigarettes, but it is not generally known that growing tobacco for export robs millions of poor people of fertile land on which they could grow cereals and vegetables. As long as there is a market for tobacco exports, agribusinesses and profit hungry farmers will plant tobacco instead of wheat, corn, or soy. The market for tobacco exports will remain as long as large numbers of people continue to smoke. Tobacco, together with other cash crops such as coffee and tea, commands a considerable portion of the world's fertile lands, yet no such produce is a true necessity.

Reducing the demand for tobacco—and for animal feed, coffee, tea, and similar cash crops—would mean a healthier life for the rich and a chance for adequate nourishment for the poor. A better pattern of land use would permit feeding eight or even ten billion people without conquering new land and engaging in risky experiments with genetically manipulated crop varieties. With today's consumption patterns, on the other hand, the world's agricultural lands can barely feed the human population. It takes only one acre of productive land to provide the average Indian's agriculture-related needs, but satisfying the needs of a typical American takes fully 12 acres. Making 12 acres of productive land available for all six billion people alive today would require two more planets the size of Earth.

Driving

According to a World Bank estimate, by the year 2010 the number of motor vehicles in use will swell to one billion. Unless there is a rapid shift to new fuel technologies—which is possible, but difficult to achieve worldwide—doubling the current motor vehicle energy requirements would double the level of smog precursors and greenhouse gases. Cars and trucks would choke the streets of third world cities and the transportation arteries of developing regions. This level of motor vehicle use is not a necessity in either the industrialized or the developing world. For goods transport, rails and rivers could be more effectively used; and for city dwellers, public transportation could be pressed into wide-scale service, reducing the number of private vehicles. In most cases this would reduce the material standard of living, but not the quality of life.

Being moral in our day means thinking twice before taking one's car to town when public transport is available. It means taking pride in clean and well-kept subways, trams, and buses, and traveling sociably in the company of others rather than in the air-conditioned and cell-phone and hi-fi equipped isolation of a private automobile.

If one is physically fit, short trips by bicycle make for a happier choice still: besides saving fuel, reducing traffic congestion, and cutting down on pollution, one benefits from an extra dose of fresh air and exercise.

We know that the urban sprawl created by the widespread use of private automobiles is undesirable, that traffic jams are frustrating and counter-productive, and that the gasoline-powered internal combustion engine uses up finite resources and contributes to air pollution and global warming. Today there are perfectly good alternatives to the classic automobile: cars running on natural gas, fuel cells, compressed air, or liquid hydrogen, to mention but a few. Yet people continue to demand and use gasoline powered cars. As long as the demand keeps up, industries will not introduce alternative fuels and cities and states will not procure cleaner and more efficient public transportation.

The switch from the liberal morality of classical industrial society to a more global and responsible ethic is slow in coming. The initially noble but now outdated liberal precept "live and let live" persists. For the most part, affluent people still live in a way that reduces the chances for the poor to achieve an acceptable quality of life. If all people used and overused private cars, smoked, ate a heavy meat diet, and used the myriad appliances that go with the affluent lifestyle, many of the essential resources of the planet would be rapidly exhausted and its self-generative powers would be drastically reduced.

Clearly, the poor people of the world must also adopt a planetary ethic. If they persist in pursuing the values and lifestyles of the affluent, little will be gained. It is not enough for well-to-do Americans, Europeans, and Japanese to reduce harmful industrial, residential, and transportation emissions and cut down on gross energy consumption. If the Chinese, the Indian, and other poor populations continue to burn coal for electricity and wood for cooking, implement classical Industrial Age economic policies, and acquire Western living and consumer habits, nothing will be gained. Only if a critical mass of people in the contemporary world adopt a planetary ethic do we have a realistic chance of creating a world where the right to life and well-being is assured for all, and the human impact on the environment does not exceed the self-regenerative capacities of the life-supporting environment.

9

Rational Sustainability

Joseph A. Tainter

Sustainability is a divisive concept. This is partially because there is a tendency to think that sustainability refers to absolute conditions of the biophysical world. Those who hold this view believe that sustainability will emerge from such things as lower consumption of goods and services, and diverting minimal energy from biophysical to human use.[1,2] Invariably this view produces conflict, for most people will not readily adopt a program that would appear to require lower consumption and less opportunity.

Sustainability conflict is often misplaced. Rather than deriving from pristine environments, sustainability is a matter of values. We sustain what we value, which for most people is an accustomed way of life, opportunities to improve one's situation, or environmental conditions that match an idealized concept of nature. Those who advocate rigorous biophysical sustainability do so because this is what they value. While environmental advocates often feel that the biological world has intrinsic value, natural systems in fact have only the values that humans impute to them. Ecosystems cannot care in which configuration they exist; and for any given ecosystem, many different configurations can be sustainable.[3] It is often argued that human sustainability depends on functioning ecosystems. Yet humans routinely extract a living from natural environments that appear to biologists to be degraded. Thus, the configurations of society or ecosystems that we try to sustain are those that we value, but therein arises the dilemma. Values are variable, mutable, and transient, which means that sustainability is always relative.[4,5]

This conflict will never disappear. We can, though, work toward making the debate rational and the discussion explicit. This would lessen the conflicts that now arise from miscommunication and unexamined assumptions. This paper addresses that goal by proposing a framework applicable across sustainability programs. The elements are (a) understanding that sustainability is relative and value laden; (b) being rigorous in defining sustainability and in monitoring for it; and (c) focusing on the contexts that produce sustainability rather than the outcomes of sustainability programs.

Relativity and Values in Sustainability

Many people involved in sustainability efforts assume that the value of the things they wish to sustain is axiomatic. Conservation biologists exemplify this most clearly, but they are not alone. Conservation biologists want to sustain biological diversity, which itself requires undegraded vegetation and landscapes. Yet to the public, land users, and legislators, the value of conserving such organisms as small fish or insects is not self-evident. Such people pay for conservation programs, and they value conservation of charismatic species: mammals, birds, and larger fish. Conservation biologists ground their arguments in science: diversity ensures proper ecosystem functioning. Yet from the perspective of energy flow, species composition is of less importance than many think. If a species is eliminated from an ecosystem, some other organism will use the energy that is liberated. System function generally continues, although perhaps in an altered form.[6] As wolves were eliminated from many North American ecosystems, for example, their place was taken by coyotes, whose range has now expanded to the eastern seaboard.

Landscapes possess tangibility that abstractions such as species diversity or energy flow lack. Yet landscapes present prime examples of the relativity of sustainability. Degradation may be the opposite of sustainability, yet it manifests itself in counterintuitive ways. Sander van der Leeuw and his colleagues have studied degradation across parts of Europe and the Mediterranean basin. Van der Leeuw points out that degradation is a social construct. It lacks absolute biophysical references. In the Vera Basin of Spain degradation appears as erosion, a common understanding of the term. Yet in Epirus, Greece, degradation is an increase of shrub vegetation that chokes the landscape. A centuries-old pastoral life, with sustainable villages, is now impossible. To urban residents the landscape appears "natural," but to Epirotes it is degraded. The spread of shrubs and trees reduces groundwater and the flow of springs. As mountain vegetation thrives, that lower down declines. When soil is eroded from the mountains it forms rich deposits in valleys that have sustained agriculture for millennia.[7,8,9,10] Erosion inhibits sustainability in the Vera Basin, but enhances it for Epirote farmers. We see in this example that in the realm of sustainability and degradation there are winners and losers. These terms mean what people need them to mean in specific circumstances.

Since sustainability involves values, its meaning varies among individuals, groups, and societies, and changes over time. Prior to the nineteenth century, the ideal landscape was agricultural. This was the landscape to be sustained, and in early America was considered the basis of Jeffersonian democracy. Today a landscape of small farmers is a largely quaint remembrance, sustained for reasons that are partly political. Many urban residents of industrialized nations today value wilderness, preferring recreation to commodity production. That value, too, may change. We should wonder why we fight political battles over sustaining forests that take centuries to mature, when centuries from now the debate may be quite different.

Directing sustainability efforts in productive directions, then, requires understanding that it is a matter of values, not invariant biophysical processes. Some people and some ecosystems benefit from sustainability efforts, while others don't. When confronted with the term "sustainability," one should always ask: sustain what, for whom, for how long, and at what cost? Only when these questions have been answered can a sustainability program begin.

Defining and Monitoring Sustainability

A definition of sustainability is essential so that those who debate it know what they disagree about. The least useful conceptions of sustainability are narrowly focused within specific disciplines. These practitioners assume that the phenomenon to be sustained has intrinsic values that should be evident to all. More sophisticated definitions respond to the value laden nature of sustainability by being general. The best known was offered in 1987 by Gro Harlem Bruntland, then Prime Minister of Norway: "Sustainable development is development that meets the needs of the present without compromising the ability of future generations to meet their own needs."[11] While Brundtland's definition will continue to be widely cited, it is not genuinely useful. Fundamentally, it borders on tautology: of course sustainable development concerns tending to the future. The term "needs" suggests material requirements, though the term could be stretched to cover the intangibles that many people in industrial nations value, such as endangered species or wilderness. The definition is too general, though, to guide decisions. As befits a political leader, it is so vague "... as to be consistent with almost any form of action (or inaction)."[12]

A different approach to sustainability has been developed by Timothy Allen, Thomas Hoekstra, and myself, built on a foundation of hierarchy, systems, and complexity theory.[13,14,15,16,17] The lessons of this work provide guidance regarding both defining sustainability and achieving it. Five lessons emerge from our analysis:[18,19]

1. Sustainability is achieved by managing for productive contexts rather than for outputs. Managing to maintain productive outputs is like sticking one's finger in a dike. Leaks spring inevitably (as output fluctuations) and must constantly be plugged. The problems generating the leaks are never addressed, so the costs of management cannot be controlled. This is how we have practiced agriculture and forestry. It is a style of management in which the integrity of the productive system may never be addressed.

2. Manage systems by managing their contexts. In a hierarchy, a system is controlled by its context.[20,21] Management efforts are best focused not on the system of interest (such as pest outbreaks in forests or agricultural fields), but on the contexts that regulate such systems (perhaps overly dense forests or monocropped fields).

3. Identify what dysfunctional systems lack and supply only that. One might apply fire, for example, to a plant community in which biomass is not decomposing as rapidly as desired. It would be inefficient to provide more than that.

4. Use ecological processes to subsidize management. In the Rio Grande gallery forest of New Mexico, for example, reforestation can be accomplished manually (and expensively) by planting

cottonwood poles. Yet spring flooding of the riparian forest (now controlled by upstream dams) would cause detritus to decay and allow a fresh generation of cottonwood seedlings to establish themselves without human intervention.[22] By restoring flooding, the forest regenerates itself through free subsidies that are available whether we use them or not: winter precipitation and gravity. The ecosystem subsidizes the human endeavor, rather than conversely.

5. Understand complexity and costliness in problem solving. As management challenges grow inexorably in scale and complexity, management grows more complex in response. Complexity is often successful in problem solving, but it can become cumbersome and costly. In southeast Alaska, for example, conflict between Native American subsistence hunting and timber production generates litigation and regulation so costly that the net value of subsistence hunting has declined.[23] Historically, problem-solving systems that develop in this way either collapse, are terminated, or come to depend on subsidies.[24,25,26] The Alaska example is expanded below.

These points lead to defining sustainability as: *maintaining, or fostering the development of, the systemic contexts that produce the goods, services, and amenities that people need or value, at an acceptable cost, for as long as they are needed or valued.*[27, 28]

This definition addresses the concern to sustain facets of the environment that we need (ecological processes) or value (wilderness). It reflects the concern to focus on productive systems rather than outputs. The four questions raised earlier (sustain what, for whom, for how long, and at what cost?) must be addressed. By focusing on values, the definition forces one to approach sustainability through the goals of a community or group. These goals may be ecosystem integrity, biological diversity, cultural diversity, recreational experiences, employment, or community continuity. Articulating the sustainability goal is the first step toward clarity, negotiation, consensus, and implementation.

Defining what one means by sustainability and what one wants to sustain are fundamental steps, but they are only the start. One needs a monitoring program to know whether sustainability goals are being met. Monitoring is also subject to values, and consistency in these values varies among disciplines and subject matter. For example, the Center for International Forestry Research (CIFOR) has studied sustainability in several tropical nations: Côte d'Ivoire, Indonesia, and Brazil. CIFOR researchers in these countries developed very different criteria for what to monitor for sustainability, and indicators for those criteria. Commonality among criteria and indicators varied greatly between these countries and across different fields. Indicators measuring ecological variables achieved 72% consistency (Figure 1), reflecting a high level of consistency in views of ecological sustainability. Indicators addressing forest management and policy achieved 60% and 57% consistency, respectively. These fields are less consistent than ecology. Social indicators proved the most variable, achieving only 34% commonality in these studies.[29,30] Concepts of social sustainability are demonstrably variable and inconsistent.

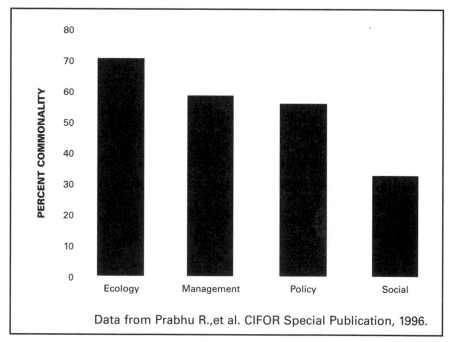

Data from Prabhu R.,et al. CIFOR Special Publication, 1996.

Figure 1. Commonality of Criteria and Indicators Across CIFOR Tests.

A large part of the inconsistency in social indicators of sustainability arises from a tendency to confuse sustainability with social improvement. This dilutes the concept of sustainability and leaves it vulnerable to political attack.[31] Sustainability is becoming a vehicle on which we load social improvements that, while they are worthwhile, bear no relation to the definitions of sustainability discussed above. Thus encumbered, sustainability is in danger of being so reduced from its original purpose that it will be difficult to distinguish it from general programs of social welfare.

Indicators of sustainability in tropical nations, for example, have included comfortable and safe transportation, health care, training, guaranteed use of roads, protection of cultural and religious sites, and quality of housing.[32] Certainly these are worthy goals, but they illustrate the problem. Such goals should be distinct from sustainability. Consider the health of tropical forest residents as a sustainability indicator.[33] Health illustrates why we should be intellectually rigorous about sustainability. It can potentially cover all of human life. Time is the test of sustainability. People who have lived in forested landscapes for generations, centuries, or millennia will be exposed to pathogens that originate there. Long-term residence in a forest environment would indicate that illnesses arising from such pathogens are not related to sustainability. The people have proven themselves sustainable notwithstanding endemic illnesses. Maladies arising from, or exacerbated by, such things as recent population growth or new forest uses, on the other hand, may be related to sustainability and thus potentially included as indicators. It is, of course, desirable to alleviate illnesses, but we should not label that a goal of sustainability. In sustainability analyses, terms like "health" and "well-being" must be broken down into those components that truly underpin sustainability and those that are desirable for other reasons.

This problem is not limited to tropical forest countries. It asserts itself in wealthy nations also. The U.S. Forest Service has also recently tested criteria and indicators of sustainable forestry.[34, 35] The social indicators to emerge from these tests include such topics as collaborative decision-making, government-to-government relationships, health and safety, and disabled access.[36] While these are all worthy goals, they are not variables that one can use to measure sustainability. Decision-making can be collaborative but ineffective. Government-to-government relationships do not necessarily reflect either sustainability or its failure. Health and disabled access are, again, worthy, but their presence or absence is unrelated to sustainability. This is the special interest group approach to sustainability. Each group clamors to have its issue included in a sustainability program. The original meaning of sustainability becomes lost in interest-group politics.

This need not be so. It is possible to develop rigorous indicators that objectively measure sustainability, such as the following:

- Is economic diversity stable, increasing, or decreasing?
- Is public infrastructure being maintained, extended, or allowed to deteriorate?
- Is the net capital flow into or out of the area?
- Is the proportion of children choosing to remain in smaller communities stable, increasing, or decreasing?

These are extracted from a set of objective indicators for rural communities, and show that sustainability can be measured free of interest-group politics.[37] Failing to do so will, in time, render sustainability monitoring irrelevant to sustainability.

Working Toward Sustainability

Sustainability is often thought, in popular discourse, to emerge as a passive consequence of consuming less. This assumes that sustainability is a plateau on which one may rest forever unchallenged. Of course this is unrealistic: the challenges that a society or ecosystem might face are, for practical purposes, endless in number and infinite in variety. This being so, sustainability is a function of problem solving. Yet problem solving is not a static capacity. The problem-solving ability of a society or smaller institution changes over time. As the problems that a society or institution faces grow in size and complexity, for example, problem solving also grows in complexity and costliness. Problem solving is an economic process, and its success depends on the ratio of the benefits of problem solving to its costs. Problem solving tends to develop from solutions that are simple and inexpensive to those that are increasingly complex and costly. Human history exemplifies the trend. Through problem solving, we have changed from simple structure and organization, egalitarian relations, and minimal economic differentiation to complex structure and organization, hierarchical relations, and great economic specialization. Each increase in complexity has costs, which may be measured in labor, time, money, or energy. Growth in complexity, like any investment, ultimately reaches diminishing returns, where increments to complexity no longer yield proportionate returns. Higher increments of complexity produce smaller increments of benefit in problem solving. When this point is reached, a society or other institution begins to experience economic weakness and disaffection of the populace, which in some ancient societies caused dramatic collapses. Thus a fundamental element of a sustainable society is that it must have sustainable institutions of problem solving. These will be institutions that give stable or increasing returns (e.g., by minimizing costs), or diminishing returns that are subsidized by energy sources of assured quality, supply, and cost.[38,39,40,41,42,43]

Ample energy sources of high quality and low cost will never be available indefinitely.[44,45,46] Thus a key element to problem solving, and to sustainability, is to expend minimum energy in the effort. In the areas of agriculture and natural resources we have, as noted, tended to manage for outputs—for specific crops such as wheat, rice, meat, or lumber. Managing for outputs is like gardening: parts must constantly be micromanaged so that they remain in the proper configuration. This is costly, and gardening is an expensive hobby. So also is problem solving that resembles gardening. Human management subsidizes the natural system, so that problem solving grows complex and costly, and ultimately ineffective. Such systems can persist only with large subsidies.

The alternative is to manage systems so that they support us, rather than conversely. My colleagues and I call this approach *supply-side sustainability*.[47] The idea is to manage the contexts of productive systems, as suggested above in the five points that led to the definition of sustainability. If we take care of the context, the system of interest manages itself and produces resources. The purpose is to minimize the costliness of problem solving. Once problem solving ceases to be gardening, the ecological system subsidizes the management effort, rather than conversely. A few examples show how these abstractions are applied in real sustainability issues.

Unsustainable Problem Solving: Conflict Over Outputs

In southeast Alaska, the Tlingit Indians try to sustain their cultural identity by hunting, fishing, and gathering traditional foods of their ancestors. These foods are essential to ceremonial life, and the hunting or gathering of them is used to socialize younger Tlingits to many facets of their culture, which is fundamental to sustaining it.

Sitka black-tailed deer is a primary game animal for Tlingit hunters. In addition to cultural value, deer meat is economically important in an area with few employment opportunities. Deer population levels depend on forage and shelter in severe winters. During snow, deer shelter and forage in high density, old growth forests, usually at lower elevations. Unfortunately, the best and most easily harvested timber is found in the same places. Thus the lands most prized for timber production are also those most valuable for deer populations and subsistence hunting.

This conflict seems to be a zero-sum game. The forests can produce economically viable levels of timber, or they can produce satisfactory numbers of deer. Politicians always want high levels of employment, so forest managers are pressured to produce timber, reducing hunting opportunities. Note that the conflict is over outputs: deer vs. trees.

The conflict between timber harvesting and subsistence hunting produces increasing complexity and costliness. In the past, a Tlingit wishing to hunt or fish could do so at will, or with the consent of the elders in another clan's territory. By 1996, six federal agencies had 74 specialists working on subsistence, with an annual budget of $11 million. In the same year, ten Federal Regional Advisory Councils had 84 members and five coordinators. Subsistence regulators hold 50 to 100 public meetings each year, and produce 50 evaluations and findings. The decision in a recent lawsuit assigned the U.S. federal government responsibility for subsistence fishing. The initial estimate of the cost to implement this mandate was an additional 250 employees and $52 million per year. In 1994 the median economic value of the subsistence fishing harvest was about $103 million. If the initial cost estimate proved correct, this one judicial decision would reduce the net economic value of subsistence fishing by 50 percent.

As the complexity and costliness of subsistence management grow, the net value of subsistence harvesting declines. Even for the Tlingit, who value traditional foods in cultural rather than economic terms, the personal cost of hunting and gathering has grown. Subsistence conflict is encumbered in a problem-solving system that is experiencing diminishing returns.[48] Even in wealthy nations, land management systems that are forced down a path of rapidly diminishing returns to problem solving cannot be sustained indefinitely. The Alaskan stalemate continues only because it is subsidized by the wealth of Alaskan oil and by the federal government.

Sustainable Problem Solving: Managing Context

Two examples illustrate the supply-side approach. The first, Rishi Valley, shows how this approach can be implemented with little more than commitment and minimal scientific knowledge. The second, fisheries management, fuses social and ecological science.

Rishi Valley of India was once a deforested landscape, in which farmers lived a marginal existence. In the 1930s, Jiddu Krishnamurti and his followers established a school in this barren place. From this small beginning a program of environmental restoration emerged that is locally-based and self-supporting. Missing contexts have been identified, and are initially provided. As these contexts assume their functions, the system restores itself.

To restore the forest one must restore its context: the landscape. Drought resistant trees and shrubs anchor soil and provide mulch; 20,000 are planted each year. Check dams slow drainage and catch nutrient-rich silt, which is transported manually to where it is needed. Stone terraces channel runoff toward large percolation tanks, which replenish the water table. Cash crops support self-sufficiency, while villagers are educated in the principles of sustainable land use. Neither irrigation nor artificial fertilizers is used. The results are that the water table has risen from twelve to as little as three meters, biomass has increased by 300 percent, and 150 species of migratory birds have returned.[49]

The key to restoration of Rishi Valley has been to use the subsidies of solar energy (producing rainfall), gravity, and dedicated people. The ecosystem both subsidizes the restoration and is enhanced by it. The subsidies minimize complexity and restoration costs, so that the effort and its benefits are localized and of appropriate scale. It appears to be a sustainable restoration.

Rishi Valley may offer lessons for urban residents in industrialized nations. Albuquerque, New Mexico, is a desert city, rapidly depleting its water table in an area that receives about 20 cm. of precipitation annually. Yet the problem may not be an imbalance between people and water, but a failure of soil management. Much land is either paved or stripped of vegetation, so that precipitation runs off. The ground surface is the context of the aquifer, and part of the solution to Albuquerque's sustainability could be to capture precipitation now lost to runoff or evaporation. Little precipitation now reaches the aquifer, but greater amounts could be captured by such practices as clustered housing, covering surfaces with porous materials, and biologically diverse open spaces.[50] Failure to manage the landscape in this way means that the city will exceed its water supply and stagnate economically.

Acheson, Wilson, and Steneck describe an approach called *parametric management* of fisheries.[51] These scientists advocate managing the ecological variables that limit population levels. Noting the failure of common stock/recruitment models to predict population levels, they advocate rules governing fishing locations, fishing area, and techniques. In this approach, one manages fish populations by managing the human influence on such factors as breeding grounds, migration routes, nursery areas, and spawning. One manages not for the outputs, as in stock/recruitment models, but for the context that affects critical processes. That context is human use. Acheson, Wilson, and Steneck observe that this is the approach taken in many tribal and peasant societies. [52] In such societies, religious sanctions and/or community pressure regulate such factors as time and location of fishing, technology, and the stage of a fish's life cycle during which it may be kept. These societies do not manage for outputs by establishing quotas. All manage the human context that affects life processes, allowing the biological system to work on its own. The human influence on reproduction may be similarly controlled in industrial societies. In the Maine lobster industry, for example, a conservation ethic in recent decades has prevented over-harvesting, so that today the lobster catch is high but apparently sustainable. The conservation measures include catching only mid-sized lobsters, and notching the tails of egg-bearing females (so that the animal is released if caught again). These practices are voluntary, but there are estimated to be millions of notched lobsters today in the Gulf of Maine.[53]

Summary and Recommendations

Sustainability will never achieve mechanistic certainty. What to sustain and how to sustain it will always generate conflict and require negotiation. Sustainability is not a goal or plateau that, once achieved, need only be monitored. Once must work constantly at sustainability. The key to sustainability is the cost of problem solving, including conflict. Sustainability efforts that are based on unexamined values generate conflict that cannot be reconciled, and can actually reduce sustainability. We cannot avoid conflict, but we can reduce its cost by making values clear, so that discussion is rational and debates are productively focused. We can reduce the cost of managing systems through the program termed supply-side sustainability, as outlined here.

Sustainability programs based on this framework:

a) accept the fact that sustainability is relative and value-based, and identify what is to be sustained;

b) define what sustainability means and determine how to monitor a sustainability trend objectively; and

c) manage for the contexts of productive systems and employ ecological subsidies.

It may seem contradictory that, while part (a) of this framework
involves recognition of values, parts (b) and (c) require intellectual
rigor and minimization of extraneous values. This is because part
(a) concerns determining what outcomes we want, while parts (b)
and (c) shift the focus to context. Context is not a matter of values;
it is what produces the things that we value. Determining context is
often a matter of scientific research. This combination of subjective
and objective concerns pervades sustainability, and contributes to
making it so awkward and contentious.

Notes

1 Rykiel, E. 2001. Scientific objectivity, value systems, and policymaking.
   BioScience 51:433–436.
2 Lackey, R. 2001. Values, policy, and ecosystem health. *BioScience*
   51:437–443.
3 Allen, T., J Tainter, & T. Hoekstra. 1999. Supply-side sustainability.
   Systems Research and *Behavioral Science* 16:403–427.
4 Tainter, J. 2001. Sustainable rural communities: general principles and
   North American indicators. In: *People Managing Forests: The Links Between
   Human Well-Being and Sustainability*. C. Colfer & Y. Byron, eds.
   Washington, D.C. and Bogor, Indonesia: Resources for the Future Press
   and Center for International Forestry Research.
5 Allen, T., J. Tainter, & T. Hoekstra. 2003. *Supply-Side Sustainability*. New
   York: Columbia University Press.
6 *Ibid.*
7 van der Leeuw, S. 1998. Introduction. In: *The Archaeomedes Project:
   Understanding the Natural and Anthropogenic Causes of Land Degradation
   and Desertification in the Mediterranean Basin*. S. van der Leeuw, ed.
   Luxembourg: Office for Official Publications of the European
   Communities.
8 Bailey, G., C. Ioakim, G. King, et al. 1998. Northwestern Epirus in the
   paleolithic. In: *The Archaeomedes Project: Understanding the Natural and
   Anthropogenic Causes of Land Degradation and Desertification in the
   Mediterranean Basin*. S. van der Leeuw, ed. Luxembourg: Office for
   Official Publications of the European Communities.
9 Bailiff, I., J. Bell, P. Castro, et al. 1998. Environmental dynamics in the
   Vera Basin. In: *The Archaeomedes Project: Understanding the Natural and
   Anthropogenic Causes of Land Degradation and Desertification in the
   Mediterranean Basin*. S. van der Leeuw, ed. Luxembourg: Office for
   Official Publications of the European Communities.

10 Green, S., G. King, V. Nitsiakos, & S. van der Leeuw. 1998. Landscape perceptions in Epirus in the late 20th century. In: *The Archaeomedes Project: Understanding the Natural and Anthropogenic Causes of Land Degradation and Desertification in the Mediterranean Basin.* S. van der Leeuw, ed. Luxembourg: Office for Official Publications of the European Communities.

11 World Commission on Environment and Development. 1987. *Our Common Future.* Oxford: Oxford University Press.

12 Pearce, D., G. Atkinson, & W. Dubourg. 1994. The economics of sustainable development. *Annual Review of Energy and Environment* 19:457–474.

13 Allen, T., J Tainter, & T. Hoekstra. 1999. Supply-side sustainability. *Systems Research and Behavioral Science* 16:403–427.

14 Allen, T., J. Tainter, & T. Hoekstra. 2003. *Supply-Side Sustainability.* New York: Columbia University Press.

15 Allen, T. & T. Starr. 1982. *Hierarchy: Perspectives for Ecological Complexity.* Chicago: University of Chicago Press.

16 Allen, T. & T. Hoekstra. 1992. *Toward A Unified Ecology.* New York: Columbia University Press.

17 Ahl, V. & T. Allen. 1996. *Hierarchy Theory: A Vision, Vocabulary, and Epistemology.* New York: Columbia University Press.

18 Tainter, J. 2001. Sustainable rural communities: general principles and North American indicators. In: *People Managing Forests: The Links Between Human Well-Being and Sustainability.* C. Colfer & Y. Byron, eds. Washington, D.C. and Bogor, Indonesia: Resources for the Future Press and Center for International Forestry Research.

19 Allen, T., J. Tainter, & T. Hoekstra. 2003. *Supply-Side Sustainability.* New York: Columbia University Press.

20 Allen, T. & T. Starr. 1982. *Hierarchy: Perspectives for Ecological Complexity.* Chicago: University of Chicago Press.

21 Allen, T. & T. Hoekstra. 1992. *Toward A Unified Ecology.* New York: Columbia University Press.

22 Crawford, C., L. Ellis, D. Shaw, & N. Umbreit. 1999. Restoration and monitoring in the middle Rio Grande bosque: current status of flood pulse related efforts. In: *Rio Grande Ecosystems: Linking Land, Water, and People.* Fort Collins, CO: USDA Forest Service, Rocky Mountain Research Station, Proceedings RMRS-P-7.

23 Tainter, J. 1997. Cultural conflict and sustainable development: managing subsistence hunting in Alaska. In: *Sustainable Development of Boreal Forests: Proceedings of the 7th International Conference of the International Boreal Forest Research Association.* Moscow: All-Russian Research and Information Center for Forest Resources.

24 Tainter, J. 1988. *The Collapse of Complex Societies.* Cambridge: Cambridge University Press.

25 Tainter, J. 1995. Sustainability of Complex Societies. *Futures* 27:397–407.

26 Tainter, J. 1996. Complexity, problem solving, and sustainable societies. In: *Getting Down to Earth: Practical Applications of Ecological Economics*. R. Costanza, O. Segura, & J. Martinez-Alier, eds. Washington, D.C.: Island Press.

27 Tainter, J. 2001. Sustainable rural communities: general principles and North American indicators. In: *People Managing Forests: The Links Between Human Well-Being and Sustainability*. C. Colfer & Y. Byron, eds. Washington, D.C. and Bogor, Indonesia: Resources for the Future Press and Center for International Forestry Research.

28 Allen, T., J. Tainter, & T. Hoekstra. 2003. *Supply-Side Sustainability*. New York: Columbia University Press.

29 Prabhu, R., C. Colfer, P. Venkateswarlu, et al. 1996. *Testing Criteria and Indicators For The Sustainable Management of Forests: Phase 1 Final Report*. Bogor, Indonesia: Center for International Forestry Research, Special Publication.

30 Tainter, J. 2001. Sustainable rural communities: general principles and North American indicators. In: *People Managing Forests: The Links Between Human Well-Being and Sustainability*. C. Colfer & Y. Byron, eds. Washington, D.C. and Bogor, Indonesia: Resources for the Future Press and Center for International Forestry Research.

31 *Ibid.*

32 Prabhu, R., C. Colfer, P. Venkateswarlu, et al. 1996. *Testing Criteria and Indicators For The Sustainable Management of Forests: Phase 1 Final Report*. Bogor, Indonesia: Center for International Forestry Research, Special Publication.

33 Colfer, C., with R. Prabhu & E. Wollenberg. 1995. *Principles, Criteria, and Indicators: Applying Occum's Razor to the People-Forestry Link*. Bogor, Indonesia: Center for International Forestry Research Working Paper 8.

34 Woodley, S., G. Alward, L. Gutierrez, et al. 2000. *North American Test of Criteria and Indicators of Sustainable Forestry*. Fort Collins, CO: USDA Forest Service, Inventory and Monitoring Institute, Report 3.

35 Wright, P., G. Alward, T. Hoekstra, et al. 2002. *Monitoring for Forest Management Unit Scale Sustainability: The Local Unit Criteria and Indicators Development (LUCID) Test*. Fort Collins, CO: USDA Forest Service, Inventory and Monitoring Institute, Report 4.

36 *Ibid.*

37 Tainter, J. 2001. Sustainable rural communities: general principles and North American indicators. In: *People Managing Forests: The Links Between Human Well-Being and Sustainability*. C. Colfer & Y. Byron, eds. Washington, D.C. and Bogor, Indonesia: Resources for the Future Press and Center for International Forestry Research.

38 Tainter, J. 1988. *The Collapse of Complex Societies*. Cambridge: Cambridge University Press.

39 Tainter, J. 1995. Sustainability of Complex Societies. *Futures* 27:397–407.

40 Tainter, J. 1996. Complexity, problem solving, and sustainable societies. In: *Getting Down to Earth: Practical Applications of Ecological Economics*. R. Costanza, O. Segura, & J. Martinez-Alier, eds. Washington, D.C.: Island Press.

41 Tainter, J. 2000. Global change, history, and sustainability. In: *The Way The Wind Blows: Climate, History, and Human Action*. R.J. McIntosh., J.A. Tainter, & S.K. McIntosh, eds. New York: Columbia University Press.

42 Tainter, J. 2000. Problem solving: complexity, history, sustainability. *Population and Environment* 22:3–41.

43 Allen, T., J. Tainter, & T. Hoekstra. 2003. *Supply-Side Sustainability*. New York: Columbia University Press.

44 Hall, C., C. Cleveland & R. Kaufmann. 1992. *Energy and Resource Quality: The Ecology of The Economic Process*. Niwot, CO: University Press of Colorado.

45 Allen, T., J. Tainter, & T. Hoekstra. 2001. Complexity, energy transformations, and post-normal science. In: *Advances in Energy Studies, Second International Workshop: Exploring Supplies, Constraints, and Strategies*. S. Ulgiati, M. Brown, M. Giampietro, eds. Padua, Italy: Servizi Grafici Editoriali.

46 Allen, T., J. Tainter, J. Pires, & T. Hoekstra. 2001. Dragnet ecology—"Just the facts, ma'am": The privilege of science in a postmodern world. *BioScience* 51:475–485.

47 Allen, T., J Tainter, & T. Hoekstra. 1999. Supply-side sustainability. *Systems Research and Behavioral Science* 16:403–427.

48 Tainter, J. 1997. Cultural conflict and sustainable development: managing subsistence hunting in Alaska. In: *Sustainable Development of Boreal Forests: Proceedings of the 7th International Conference of the International Boreal Forest Research Association*. Moscow: All-Russian Research and Information Center for Forest Resources.

49 Kaplan, R. 1996. *The Ends of the Earth: A Journey At the Dawn of the 21st Century*. New York: Random House.

50 Horst, S. Slipping through our fingers. *Albuquerque Tribune* July 28, 2001:C1–C2.

51 Acheson, J., J. Wilson, & R. Steneck. 1998. Managing chaotic fisheries. In: *Linking Social and Ecological Systems: Management Practices and Social Mechanisms for Building Resilience*. F. Berkes & C. Folke, eds. Cambridge: Cambridge University Press.

52 *Ibid.*

53 *Ibid.*

10

Seeing the Whole Picture

Richard B. Norgaard & Paul Baer*

Wl e live in a highly interconnected world. The behavior of each of us impacts on all of us, though in complex ways. If we do not acknowledge these interconnections, we will heat the globe, diminish biodiversity, and accumulate toxic materials to our demise. Our survival depends on our ability to see and respond to the complex ways we are interconnected. Our knowledge, however, is concentrated around disciplines with separate and narrow gazes on the interconnections of the world, like separate binoculars on fixed mounts. In theory, as the disciplines advance, the field of vision of each discipline widens until the combination of disciplinary binoculars looks out upon the whole picture. But even if this description of the process, rooted in the idea of the unity of science, had some basis in practice, how would the images from the separate binoculars actually be put together? Who would look through all of the binoculars and see the

* This chapter was originally drafted as a discussion document in support of ongoing research funded by the U.S. National Science Foundation Biocomplexity Program (NSF SES-0119875).

whole picture? How would he or she convey the picture to us? And on what basis would the rest of us agree to the information being relayed to us by this special observer of the whole? Obviously there can be no such special person. We must somehow do it ourselves.

We argue in this chapter that how we think we will understand across disciplines to see interconnections is poorly developed within science and among the populace at large. Historic understandings rooted in the belief that the sciences would simply unify have served us poorly, while the actual ways we do learn across disciplines are poorly recognized. The survival research we propose addresses this shortcoming. This essay lays out the issues.

Establishing institutions so that our individual activities maximize the well-being of all requires, at a minimum, a broad collective acceptance of how we understand across our disciplinary knowledge. This simple statement hides serious issues. Who are the "we" who understand, how does this "we" understand across the separate disciplinary territories of the natural and social sciences, and how collective must this understanding be to work with democratic governance? Our emphasis is on the different ways we understand across the disciplines, and how these might be improved.

Let's elaborate an example—not an entirely happy one, but a case that illustrates the key issues. Scientists undertaking research on global climate change have accumulated understanding through a variety of individual, interactive, and integrated research efforts. Atmospheric chemists are working in terms of milliseconds and microns, evolutionary ecologists at the scale of continents and millennia. Big and little computer models, lab bench and field studies, and scenario building and environmental valuation are all important. Through diverse analytical and synthetic processes, scientists addressing global climate change have come to a collective understanding, expressed in the executive summaries of the reports of the Intergovernmental Panel on Climate Change. In their judgment, people need to significantly alter their relations with the environment. The scientists within this community are not of one mind. Quite the contrary; a variety of scientific judgments exist, with some of the differences arrayed along disciplinary lines. Some of the economists participating have voiced judgments of the overall situation that are less dire, while some environmental scientists concerned with the fate of particular populations, including human, are concerned that the situation is extremely serious.

At the same time, some scientists who are largely outside of the community of global change research are casting doubt on how those within the community understand the systemic process of global climate change. Critics argue that the science is too incomplete and uncertain to frame a policy response. They accuse climate change scientists of basing their policy recommendations on their collective scientific judgment rather than the science itself. Furthermore, some critics argue that the scientists are a self-selected community predisposed to think climate change is important. The controversy, in part instigated and sustained by corporations with an interest in maintaining a fossil-fuel-based economy, has confused the public and politicians in the United States. Many now feel that it is only rational to delay action until the science is sufficiently strong.[†][1] Thus we have a problem of how we understand and how we translate understanding into action that threatens human survival as we know it.[2, 3]

The politics of global survival is largely a game of individual and corporate greed played on a field of human short-sightedness. The particular political maneuvers in the debate over climate change, however, also clearly play off historical understandings of how science is supposed to work and to connect to the policy process. To some extent, powerful interests are misconstruing appropriate principles about how science works. At the same time, scientists and philosophers have not adequately addressed how we understand across disciplines. Some of the ways we understand across disciplines are not a part of public understanding, or even scientific understanding, of how we know. A very simple example should suffice. With only one globe on which to experiment, for example, we cannot insist on model validation through experimental testing. The best we can do is see how well a model fits quite limited historical data, with human activity being insignificant for most of the record. Seeing environmental problems as systemic problems between social and environmental systems, especially when the problems are also global, forces us to rethink how we know. This is the realm of epistemology, the branch of philosophy that addresses how we think we know. Establishing deeper and richer understandings of how science works would reduce the opportunities for corporate greed to trump the understanding of scientists.

---

† The arguments and tactics of those trying to discredit the science of climage change have evolved over time. Ross Gelbspan (1997) provides an excellent lay summary of their efforts into the mid-1990s in his *The Heat Is On: The High States Battle Over Earth's Threatened Climate.*

Four Approaches to Understanding Across the Disciplines

There seem to be four different ways that we understand, or think we understand, across disciplines. We use all four approaches, and have for some time, but only what we refer to as the "unity of science approach" is widely held by scientists and generally a part of public consciousness. To some extent the different approaches complement each other through different strengths and combine to form interesting hybrids. By explicating how these approaches differ yet complement each other, a clearer view of the epistemological challenges to global predicaments emerges.

A further caveat is in order before we proceed. While we see these four approaches as central to advancing how we understand whole systems, it is important to note that most scientists, and certainly the public, pay little attention to any of them. By default, the dominant approach is to not ask questions about interconnections. When questions are raised, historic beliefs about divisibility of systems are appealed to, rationales about small changes on the margin not affecting other parts of the system are invoked, or responsibilities for interconnecting effects are simply shrugged off.

The Unity of Science.

The dominant explanation of how we understand across disciplines relies in a long-standing belief in the unity of science.[4] Science has progressed by a process of separation, by having physicists, chemists, geologists, zoologists, botanists, and scientists of many additional disciplines investigating the inner workings of separate parts of reality. There is, of course, some consternation about how this process that is taking place across the whole of science will actually yield knowledge that is in any sense whole.[‡] Nevertheless, we still refer to our

---

[‡] The social sciences seem especially problematic. At least in the physical sciences, all college students start out with the same basic courses in physics and chemistry. One introductory text and course serves all of the biological sciences. Though this is proving increasingly tenuous, historically, a single course in western civilization provided the historical and philosophical bases for all of the humanities and social sciences. This began to disappear by mid-20th century and had almost completely disappeared by the end of the century. Students enter the individual social science disciplines without a synthetic introduction to the social sciences as a whole. The individual disciplines not only exist separately but have broken into competing schools of thought with respect to the

schools of higher learning as universities rather than multiversity's because we believe that as the separate disciplines continue to advance, they will inevitably merge into a coherent whole with all of the interconnections in place. After all, the scientists of diverse disciplines are all studying one reality. If what they are learning is true, then as the knowledge acquired in each discipline expands and improves in quality, it will surely merge with the knowledge acquired in other disciplines. From this perspective, when faced with incomplete understanding of the connections between disciplines, it is logical to argue that we need still more research on the separate parts until the connections become apparent.

Many argue that our survival necessitates a significant reduction in the gaps and contradictions between what ecologists and economists think they know. The unity of science view suggests that we ought to be pushing economics toward ecology, and *vice versa*.§[5] In the naive view of the unity of science, a bridge can be made simply by directing existing research methods in the individual disciplines toward each other. The disciplines will eventually "naturally" join, such that the whole ecological-economic system necessary for resolving any particular predicament is understood. There is at least some historical evidence that this approach works. While the physical and microbiological sciences are far from fully merged, biophysics and biochemistry provide fairly wide bridges. Yet, though the laws of thermodynamics hold across the natural sciences, we also know that basic principles from the physical sciences do not explain life, that our knowledge of biological organisms does not explain organization at the ecological level. New principles must be introduced at different levels of organization.

Thus more sophisticated discussions of the ways in which the sciences could possibly, or not, be unified pays attention to issues of

---

same issues, and these divisions seemed to have established long run equilibria.

§ Economists and ecologists joined to form the International Society for Ecological Economics in 1988 largely on this rationale. It was apparent from the beginning, however, that ecology and economics each had many models. Some of these were of the same structure. Economic models of markets merge with ecological models of population interaction: input-output models in economics can be expanded to include food webs; and energetics models can be applied across both systems. Others, however, simply had no counterpart in the other discipline.

language, foci, frameworks, and methods embedded in how we do science. Can the languages of the sciences really be unified? Can, or must, all phenomena be reduced to physical objects so that all explanations build up from physics? Or is unity to be found in a limited number of patterns and dynamical relationships? And can it be argued that a single method would unify science, or does the existing plurality of methods assure disunity?**6

The proposition that the disciplines will eventually merge is the dominant understanding of how we will eventually know across the disciplines. This belief is consistent with the existence of disciplines and the structure of our universities. It is compatible with progressive governance where the management of air, water, soil, wildlife, energy, and fissionable materials, for example, are handled by separate agencies and subagencies. It justifies why those in dominant schools of thought within disciplines "should" dismiss other schools of thought as evolutionary relics. This belief girds most scholars' understanding of what it means to do interdisciplinary research. In short, it underlies and therefore does not threaten the existing dominant institutional structures of science and its application. Compatibility with existing institutions gives the unity of science vision a strong advantage over other beliefs with respect to how we understand across disciplines and with respect to how science leads to collective policy and action.

At the same time, the unity of science vision has been dominant during several critical centuries and has failed us. It has underlain how we thought we would find interconnections between disciplinary knowledges while we worked on specific problems, yet cross-cutting problems have become worse, are becoming global in the process, and now threaten our future. There seem to be several critical reasons why science is not unifying adequately to meet our needs.

** Putting all knowledge in a well cross-referenced encyclopedia using common language has long been portrayed as a part of the process of unification and was the primary motive of the efforts of Rudolf Carnap, Otto Neurath and Charles Morris to assemble an encyclopedia of unified science. Metaphors of unification also abound. C.S. Peirce argued that the diverse separate short strands of knowledge could be wound together to be strong and functional like a rope. Neurath used the metaphor of an orchestra of separate musicians playing different instruments. To get beyond the question of who wrote the score for the orchestra and who will conduct it, Norgaard (2001) proposed the term "improvisation" of "discordant knowledges" over the orchestration of complementary disciplinary instruments.

First, scientists within disciplines have become increasingly isolated and have not updated—through keeping up with developments in other disciplines—how their separate knowledge fits into the whole picture. This raises serious questions about how the unity of science actually unifies across multiple minds. And even if it could be unified within science, one has to ask how the knowledge inherent in the unity can be drawn upon by the public at large. This problem manifests itself in different forms in the three additional approaches we identify for understanding interconnections, so it is perhaps the most difficult problem.

Second, what could have been and ideally should be a conscious social process toward unity in assumptions across the disciplines based on shared learning has actually been going in the opposite direction. Outdated explanations as to how any individual scientist's particular work fits into the whole have become even more strongly reinforced within increasingly separate disciplinary cultures and have become part of the shared beliefs that help bind disciplinary participants together as a culture. Ironically, if these shared beliefs about how the disciplines connect were not outdated, i.e., if they were constantly changing in keeping with developments in knowledge in other disciplines, they could not constitute a tradition around which a disciplinary culture coheres through time.[††][7, 8]

Third, while some patterns of thinking, or presumptions about the nature of systems, are the same across disciplines, there are different patterns of thinking, even within disciplines, that are truly different and will never merge. Newton's mechanical explanation and Darwin's evolutionary explanation of change are used across multiple disciplines. Different patterns of thinking can be used iteratively together, but they will not merge. There is not a meta model that explains both of the processes. They are simply different and do not cohere. Without coherence, scientists go into the policy process with multiple patterns of explanations that are quite likely to yield multiple insights.

This does not mean there is no hope for further unification of the sciences. Further merging of disciplinary knowledge is likely and will prove helpful to an understanding of connections between

---

[††] Norgaard (2002) posits that disciplinary distancing through the maintenance of assumptions rooted in out-dated knowledge helps explain why the arguments presented in Bjorn Lomborg's *The Skeptical Environmentalist* (2001) were well received by many social scientists while they were so divergent from the understanding of environmental scientists.

the disciplines. The failings of this approach, however, indicate that we also need to pay attention to other approaches to understanding across the disciplines. Three additional approaches are presented in Table 1. While beliefs in the unity of knowledge, the upper left hand box, are widely held, the other three approaches to it are rarely explicitly discussed. For this reason, we will elaborate on each of the three alternative approaches in some detail.

Unity of Science: The problems of complexity and understanding of whole systems will diminish as the disciplines expand their understanding and eventually merge into a coherent explanation of reality.	**Integrated Assessment:** We can construct systematic ways and professional standards for integrating models and information developed in the disciplines into larger models that generate information about larger complex systems over appropriate time periods. The problems of complexity will go away as we develop and put our trust in professionals trained to undertake integrated assessment modeling.
Heuristic Models: All models ignore great complexity at smaller/lower/shorter scales. All problems can be modeled fairly simply on the scales at which we need to understand them. The problems of complexity and understanding whole systems will be reduced as a new breed of scientists arises that develops and uses models, and develops data sets, at appropriate scales.	**Distributed Learning Networks:** While methodological pluralism is accepted, different scientists using different models and languages focusing on different parts of the system can communicate their findings to each other, affect each other's research design and their interpretation of results, and can make collective assessments of the whole system. The problems of complexity will be reduced by deliberately designing distributed learning networks to make them work better and learning how to draw upon, and learning how to trust, collective assessments from the networks.

Table 1. Four Modes for Understanding Across the Disciplines.

Integrated Assessment.

There is a long history of scientists being asked by legislators to make their best assessment of a current situation and the outlook for the future. Frequently these assessments occur within or between governmental agencies using their own scientists. Scientists from academe, however, also participate in such assessments individually, through science advisory boards, and through their national academies of science or similar organizations. To this informal process of assessment, economists introduced the formal tools of linear programming and cost-effectiveness analysis during World War II, and then cost-benefit analysis, largely during the 1950s. Economic valuation was extended to environmental and social costs and benefits that were not monetized in markets.[9, 10] By the 1960s, computers began to provide ways of looking at complex data sets, giving rise to multi-criteria analysis. In the 1970s, proposals for projects affecting the environment had to be accompanied by environmental assessments, and these greatly expanded the range of questions being addressed. By the 1980s, computers were being used increasingly to link dynamic models. Now many scientists and engineers are involved in assessment as their sole activity. They follow different approaches and work at different levels of complexity and integration based on their training and how the historical demands for assessments have shaped their experience. Professional consulting corporations also now play an important role in the process.

The term "integrated assessment," or IA, traces to the interdisciplinary project to evaluate the possible global impacts of supersonic jet transport in the 1970s. The term has been invoked in a variety of ways since. The diverse information typical of environmental impact assessments of projects can be assembled in an integrated manner. Furthermore, such information can be subjected to a systematic multi-criteria analysis. We use the term, however, as it is being used today with respect to global problems where it refers to projects where quantitative models from multiple disciplines are linked together for the purposes of developing and evaluating the economic, social, and environmental outcomes of alternative starting assumptions and policy options. Integrated assessment modeling in this sense arose with the energy model developed at the International Institute for Applied Systems Analysis in the 1980s. It became a key part of the effort to develop climate change scenarios and understand the effectiveness of different policy options in the 1990s. The term "integrated assessment modeling" is more specific, but we will use the more general term since models are integral to all four of the broad processes by which we understand across disciplines.

Integrated assessment (IA) claims to integrate the models of specific disciplines into a more comprehensive model. The premise of IA is that the whole system can be understood as being made up of subsystems that can be taken from the disciplines as closed independent modules, and then linked at their boundaries. Furthermore, the linkages between the models result in feedbacks on the models from one time period to the next. The original subsystem models, because they are largely coming from the disciplines, have been developed pretty much independently of each other and may run at different spatial and temporal scales. Both to get the modules to work together and because of data limitations, IA experts may have to modify the original models, build simpler models that incorporate critical elements, or provide an interlinking model that aggregates output across space and time. Integration requires that one or more critical outputs of each module must be inputs to one or more other modules. The flows between the modules must necessarily be quantitative in nature.[11]

IA modeling piggybacks on the broader "unity of science" bandwagon. After all, if each of the disciplines is independently developing predictive theory within its own domain, then combining the disciplinary models into a single integrated model must give the overall best prediction. If it turns out that particular connections between the disciplines are left out of the IA model today, the model can be improved by adding those connections at a later date.

It is difficult to describe integrated assessment without dwelling on its many targets of controversy. Models developed and heretofore interpreted within individual disciplines are being taken out of the hands of those in the disciplines, being modified, and being used with other models in ways over which the disciplines have no control. Data are being selected and outputs managed to keep the integration from exploding apart so that IA provides "plausible" outputs. It is probably better to think of those who practice IA as professional experts rather than as scientists as we have thought of them historically. At the same time, this new breed of professional experts are being asked to "speak to" what have been posed as scientific questions while their answers "speak for" scientists as a whole. IA draws on the credibility of the models and broader knowledge developed in the disciplines. It uses, transforms, and augments the languages of the sciences as a whole. Linkages are also being made the other way, as individuals from the disciplines draw on IA both to interpret what is needed from the disciplines and to justify their work. In the process, what we understand to be science continues to evolve, though few see the process as a clear case of progress.

Distributed Learning Networks

We argued that the disciplines seem to be growing apart rather than together in the discussion of the problems with putting our faith in the eventual unity of science. At the same time, there are important instances of scientists coping with the disunity of science. We argue in this section that a key approach has been a process of distributed learning among scientists using different approaches working in different fields. If we can rationalize and improve upon the social processes we have used to cope with disunity, science may be able to better address the interconnections between the parts studied within the disciplines. Two examples are illustrative.[12, 13, 14, 15]

Our understanding of modern agriculture combines the ways soil scientists, agronomists, entomologists, plant pathologists, agricultural economists, and the other agricultural scientists understand agriculture systems. Scientists from different disciplines sometimes work together on particular agricultural problems. Yet these separate disciplines have not combined; there is no over-arching meta model. Rather, scientists from multiple disciplines have been talking to each other for over a century. They are clustered in colleges of agriculture and participate in regional, national, and international meetings. Any scientific description of modern agriculture will dwell heavily on the shared understandings these scientists have developed about the agricultural system. To some extent, the shared understandings have coevolved around specific collective efforts to formally assess the strengths and weaknesses of modern agriculture, though it is equally true that such deliberate "new" assessments strongly reflect historically shared understandings. These understandings are tested in practice and modified when a sufficient number of modern agricultural scientists acknowledge that the practices are inappropriate. Not infrequently, the decision is carried by the weight of opinion held by scientists outside of the particular discipline that initially promoted the practice. The systemic problems—ecological, economic, and social—of modern agriculture indicate that the shared understandings and the social processes through which the understandings change have some significant failings. In spite of the obvious problems, and perhaps even because of them, the ways and extent to which modern agricultural scientists know across their disciplines presents an interesting case study.

The Intergovernmental Panel on Climate Change is also a large group of scientists from different disciplines working in fairly close communication. The big computer models of our global climate do not link the knowledge of the atmospheric physicist working at the scale of milliseconds and microns to the knowledge of the evolutionary biologist working at the scale of millennia and continents. As an interactive group of researchers, however, advances made in one discipline affect how researchers in other disciplines design their subsequent research projects. Think of them as a network of interactive researchers, or a distributed learning system. Among them they have systematic knowledge that is probably improving as they learn with each other. They design their research in the context of being a member of a distributed learning system, yet no one of them understands all of the connections between their disciplines. Collectively, they are the best informed people we have on the subject, but they do not have a coherent model that predicts specific futures associated with specific policies. What they know is affected by the social processes of the climate change scientific community, the formal and informal communication channels set up by the IPCC and half a dozen other key institutions, and the scientific journals through which they formally communicate. They are the best informed, so surely when they provide us with their collective judgment as to what we should do, we should take them seriously.‡‡

These two examples, when presented as reasonably successful cases of understanding across disciplines rather than as problematic cases of coping with the disunity of science, provide some interesting insights. Methodological pluralists hold that science consists of multiple different simple models that cannot come together into a coherent whole because they are indeed different. Specific models can be made more complex, but that can also be

---

‡‡ Note that we are neither arguing that modern agricultural scientists or global change scientists were consciously organized as a distributed learning network nor that scientists have consciously participated in a distributed learning network.

self defeating, as argued above with respect to the problems of integrated assessment. Models are simplifications of reality, and it is because they are simplifications that they help us. A map with all the detail of reality would be unmanageable. Without coherence coming through convergence—as in the unity of knowledge vision or the judgment of professionals engaged in integrated assessment—understanding across disciplines can still arise through a social learning and collective judgment process. While the way we gain understanding across disciplines has long incorporated such processes, there has been relatively little work on how different social structures, communication rules, and judgment processes hinder or help scientists in reaching improved understanding across the disciplines.

A key problem of this approach is that the dominant expectation of how policy follows from science does not rest on collective judgment. The dominant expectation is that there is a single coherent model, or perhaps multiple complementary models, that predict alternative futures reasonably accurately. Collective judgment seems too subjective a process. The IPCC scientists, for example, have accepted public funds to scientifically determine the effects of climate change. They have not been asked to prepare an "assessment" in the "best judgment" manner used, for example, by the Paley Commission to assess natural resource availability in the United States after World War II. Why is the IPCC giving us their best judgment rather than proven scientific results? And what is the process by which this collective judgment was reached? All of the scientists did not vote on each sentence in the final report. As experts on only parts of the system, each would be obligated as a scientist to excuse himself or herself from passing scientific judgment on the vast majority of the report.[16] These issues indicate that the process of learning across disciplines through distributed learning networks needs to be improved, formalized, taught to scientists, and disseminated to the public.

Heuristic Models.

A fourth way that we understand across disciplines is easy to describe, difficult to practice, and, indeed, fairly rare. We tend to think of systems being made up of the multiple parts we commonly model. Complexity arises as we try to build up from simple models to the complexities of the system as a whole. Our fourth approach starts with the argument that all models are heuristic, or simply tools for discovery, not representations of reality itself. We tend to think of our models as representing a simplified reality, but think of that simplification as a matter of reducing the complexity of the larger system. We then think of understanding the complexity of larger systems as a matter of building up to the whole system, much as is done in integrated assessment. In fact, however, the models we typically use ignore complexity at lower levels much as they do at upper levels. For example, a population biology model of the interactions of a predator and a prey species simplifies the larger reality by ignoring interactions between the predator and prey and a multitude of other species. Ecologists have tried to build "up" from such simple models by bringing additional species interactions into the models to understand ecosystems. So we think of the complexity of interconnections between disciplines as being associated with building up. But it is also true that the predator-prey model is simple because it ignores the complexity that is known by organismal biologists and microbiologists, to say nothing of the complexities addressed by chemists and physicists. At the same time, it helps to recognize that the models we need to understand global climate change do not have to include, for example, the "higher" complexities being studied by astrophysicists concerned with the cosmology of the universe. All models define their system within some focus or boundary; we simply need models that focus on the boundaries appropriate to the human problem under consideration.

One of the best examples of a well focused, simple model is the infamous 'Nordhaus model' of climate change.[17] In merely 13 equations, the model provides an economic production function, including technological change, of how greenhouse gas emissions stem from production; how gas emissions accumulate and cause climate change, and are ameliorated by sequestration in the ocean as a function of ocean temperatures; how climate change damages the economy; the cost of reducing greenhouse gas emissions; and a social welfare criterion to be optimized. This has been a highly influential model because it really does incorporate the critical components of multiple disciplines related to climate change. The model is transparent enough that both its shortcomings and how they affect its output can be clearly identified and easily discussed.

Acknowledging that models can be built at any scale raises several issues, many of which show up in the Nordhaus model. First, it helps us realize that the vast majority of disciplinary science for centuries have been conducted at spatial and temporal scales that now seem inappropriate for addressing global, long term problems.§§ The problem is not simply that science has fractionated into disciplines, but that scientists maintained traditions of working at scales that were "doable" and relevant to "human" problems as they were perceived at the time. Second, it helps us see that working at a new scale requires a solid scientific community accustomed to working at that scale who also write textbooks and train graduate students. Third, it helps us see that data may need to be collected at a scale that fit new models. Our understanding of global systems, necessary for survival, is limited by all three of these problems.

Further Discussion of Epistemological Issues

We noted at the beginning of the last section that the vast majority of scientists, and certainly the public, give little thought to how we understand across disciplines, arguing that this constitutes a default approach that in some sense constitutes a "first" approach. We then discussed four more constructive approaches to understanding across disciplines. It is probably also appropriate now to note a sixth approach. Some are aware of the importance of interconnections, concerned about our vulnerability but largely skeptical about the prospects that we will ever know much about the whole. This sixth mode could slow learning through the four constructive approaches to understanding across the disciplines, but it also has the policy implication that we should be more humble. Many following this approach invoke the precautionary principle, arguing for fewer and smaller changes and only when necessary. Like the other approaches, this one need not be undertaken alone. It can be blended into an overall approach.

There are several issues that cut across all four approaches. First, an overarching question is: how do we determine whether an IA or understanding developed through a distributed learning network or a heuristic model is appropriate, or one better than another? This is an age-old question in epistemology that was quieted, if not

---

§§ Clearly, physicists have long worked across scales, from subatomic particles to the universe as a whole. Economists have long addressed the macroeconomics of business cycles and depressions, and a solid community of economists work at this scale, but a satisfactory macroeconomics still eludes them.

resolved, during the 20th century by the potential for falsification. Falsification across disciplines, however, is more difficult *a priori*. Historic data may be used, but historic data will not include how people are currently affecting multiple aspects of the system. Second, while we only seriously raised the issues of temporal and spatial scale with respect to the use of heuristic models, the other three approaches assume that scientific knowledge garnered working at historic scales will somehow cohere, be integrated, or networked into understanding systems at the larger spatial and temporal scales necessary for our survival. With so few scientists working on whole systems, let alone at global scales, it is no wonder that so many are uncomfortable with the global understanding of those who are trying to do so. Third, advancing any and all of the four approaches will require many more scientists who are comfortable with the strengths and weaknesses of each approach, and who are consciously practicing one or more approaches. This will require the development and acceptance of new standards of how science works and of what it means to be a scientist. Fourth, all four approaches raise questions with respect to who understands across the disciplines and how that understanding can be conveyed to the public so that it can be collectively acted upon. The disciplines may merge to a unified understanding, but who actually holds that understanding, how would we know it if someone claimed he or she did, how would we know whether he or she is right, and how can the public draw upon the understanding of one or more people? The problem is analogous for a distributed learning network. The integrated assessment and the heuristic model approach rely on the development of new classes of experts, but integration across expertise and trust in experts in democratic societies have already proven problematic.***[18] Fifth, at several points along the way we have raised questions about the relationship between the different approaches to understanding across disciplines and how the policy process is set up.

One especially critical cross cutting issue, however, has not been raised at all. While early philosophers pondered cataclysmic events, the early achievements in explaining the motions of the planets seem to have tethered the disciplines of modern science to the framing of systems as changing in regular, smooth patterns. Many

---

***For a popular reference, see Sheldon Ramptom and John Stauger *Trust Us, We're Experts* (2001).

have argued that Newton's mechanics set expectations about how systems work that we are now only beginning to overcome.[19,20]

Practical Implications

We have sorted the ways we now think we understand across the disciplines into four categories. Sorting into four categories helps us see that different ways entail different epistemological problems. For any given situation, multiple ways are typically mixed together, folding epistemological issues into each other. So our typology helped untangle the issues. Yet we did not untangle any epistemological quandaries through our typology. It simply has allowed us to identify separate quandaries more closely. Yet this could be quite constructive.

Scientists and lay people regularly live with the epistemological quandary of whether we can ever induce a general finding from multiple particular observations. It is an issue that is raised at various times in the training of most scientists, it must always be addressed, and the status of such knowledge is kept in mind, but scientists do not waste a lot of time debating over how many observations there must be. Scientists and society at large live with the quandary of induction reasonably constructively. Such, however, is not the case with the quandaries we have identified in this paper. Quite the contrary, most of the quandaries of understanding across disciplines are not recognized at all, let alone taught to scientists in graduate school. To the extent they are recognized, scientists and society are in the early stages of epistemological sophistication, fretting over them rather than the later stages of constructively working with them.

The epistemological issues related to how we understand across disciplines need to be elaborated much further so that we can move beyond the fretting and work constructively. The quandaries need to be better identified and built into graduate training generally, for many scientists end up participating in issues where understanding across the disciplines proves necessary. More sophisticated training in epistemology needs to be incorporated in interdisciplinary environmental science programs at the undergraduate and graduate levels, much like environmental ethics is today.

A more sophisticated understanding of how we understand across disciplines could be used to improve scientific projects. The Intergovernmental Panel on Climate Change, for example, appears to have been designed and operated predominantly under the belief that the application of the disciplines to the problem of climate change would result in information that would all fit together in the end. While the climate change community of scientists clearly operated as a distributed learning network, leaders and participants appear to be only marginally aware of this process rather than consciously taking advantage of it. It appears that many participants were frustrated by unrealistic expectations as the project moved toward an integrated assessment mode of learning. Lastly, the project is vulnerable to inappropriate criticisms from scientists outside the community because leaders and participants, to say nothing of the public at large, were not as cognizant as they could be of the epistemological quandaries of understanding across the disciplines.

Conclusions

Human survival will depend on scientists and the public at large seeing the whole picture. We probably have most of the pieces of the whole picture already in the form of disciplinary knowledge, or the pieces could be seen within the disciplines once the connections between the disciplines are made appropriately. Thus, we are arguing that building understanding of the interconnections across the disciplines is a critical element of seeing the whole picture. We argue that the dominant way we think we understand across the disciplines, the belief that the disciplines will naturally merge, is both significantly flawed and has prevented us from effectively developing the other ways we learn across disciplinary boundaries, through integrated assessment, heuristic models, and distributed learning networks. Scientists earn "doctorates of philosophy" early in their careers, and those who go into universities award them to others. Most scientists, however, only philosophize about basic epistemological issues during informal discussions at best. Interdisciplinary environmental programs have courses in environmental ethics, one branch of philosophy, but rarely ever address interdisciplinary epistemology. We hope we have contributed to the case for more seriously addressing the underlying epistemological issues of our current practices for understanding across disciplines, for making these issues a core part of interdisciplinary training, and for framing and exercising understanding across the disciplines more systematically in the future.

Notes

1 Gelbspan, Ross. 1997. *The Heat Is On: The High Stakes Battle Over Earth's Threatened Climate.* Reading MA: Addison-Wesley.

2 Jasanoff, S. & B. Wynne. 1998. Science and decision making. In: *Human Choice and Climate Change.* Vol 1. S. Rayner & E. L. Malone, eds. Columbus, Ohio: Battelle Press.

3 Miller, C. A. & P. N. Edwards, eds. 2001. *Changing the Atmosphere: Expert Knowledge and Environmental Governance.* Cambridge, MA: MIT Press.

4 Wilson, E. O. 1998. *Consilience: The Unity of Knowledge.* New York: Knopf.

5 Norgaard, R. 1989 . The case for methodological pluralism. *Ecological Economics* 1:37–57.

6 Norgaard, R.B. 2001. The improvisation of discordant knowledges. In: *The Nature of Economics and The Economics of Nature.* C. J. Cleveland, R. Costanza & D. I. Stern, eds. Cheltenham, U.K: Edward Elgar.

7 Norgaard, R.B. 2002. Optimism, pessimism, and science. *BioScience* 52(3):287–92.

8 Lomborg, B. 2001. *The Skeptical Environmentalist: Measuring the Real State of The World.* Cambridge, UK: Cambridge University Press.

9 Hanley, H. & C. Splash. 1993. *Cost-Benefit Analysis and the Environment.* Cheltenham, U.K: Edward Elgar.

10 Foster, J., ed. 1997. *Valuing Nature: Economics, Ethics, and Environment.* London: Routledge.

11 Rotmans, J. & H. Dowlatabadi. 1998. Integrated assessment modeling. In: *Human Choice and Climate Change,* Vol 3. S. Rayner & E.L. Malone, eds. Columbus, Ohio: Battelle Press.

12 Innes, J. & D. Booher. 1999. *Planning Institutions In The Network Society: Theory for Collaborative Planning.* Working Paper, Institute for Urban and Regional Planning, University of California at Berkeley. WP 1999 05.

13 Kitcher, P. 2001. *Science, Truth, and Democracy.* Oxford: Oxford University Press.

14 Dryzek, J. S. 1990. *Discursive Democracy: Politics, Policy, and Political Science.* Cambridge: Cambridge University Press.

15 Dryzek, J. S. 2000. *Deliberative Democracy and Beyond: Liberals, Critics, Contestations.* Oxford: Oxford University Press.

16 *Coal Leader.* 2001. Coal in the news: Europe's support of Kyoto. October: p. 3.

17 Nordhaus, W. J. 1994. *Managing the Global Commons: The Economics of Climate Change.* Cambridge, MA: MIT Press.

18 Rampton, S. & J. Stauber. 2000. *Trust Us, We're Experts.* New York: Penguin/Putnam

19 Norgaard, R.B. 1994. *Development Betrayed: The End of Progress and A Coevolutionary Revisioning of The Future.* London: Routledge.

20 Gleick, James. 1987. Chaos: *Making A New Science.* New York: Penguin Books.

11

What Can the Systems Community Contribute to Ensure the Survival of Civilization?

Kenneth E.F. Watt

Is It Extreme to Question the Survival of Civilization?

It may seem like blatant nonsense to suggest that it is even worth-while to consider the survival of civilization. In fact, there are many arguments based on fact, logic, reasoning by analogy from other civilizations, or studying the dynamic behavior of systems to make one very alarmed. Further, several hundred important books have appeared in the last few years suggesting that we are living in very precarious times. The most alarming fact: most of these books assert that the United States is rocketing towards catastrophe because of one particular phenomenon or process: the debt burden, global fuel shortages, the likelihood of world war over diminishing resource availability, the costs of fighting crime and terrorism, the inadequacy of the U.S. educational system, etc. Most authors are aware of only a subset of these problems. However, from a "big picture" perspective, interactions between the different societal breakdown mechanisms will bring catastrophe faster—and of much greater intensity—than most people imagine. These books are interesting because of the data and logic used in the arguments, the remarkably diverse backgrounds of the authors, and the severity of the situation as they see it. For example, both Lester Thurow and Morris Berman write explicitly about the possibility of a return to The Dark Ages.

We Are Involved in a War of Ideas

As the editor in chief of a new encyclopedia who has had great difficulty getting the work published, even by the publisher that commissioned it and made an advance payment on expenses, I am well aware of the politics swirling around new ideas. I offer a brief description of five root sources of confusion in understanding current systemic societal problems. The bulk of this article will be devoted to explaining how the systems community can deal with these.

1. The importance of putting on a happy face, and psychological manipulation of the electorate.

Last night I watched two network talk show hosts, Chris Matthews and Tim Russert, talking about the "Happy Face" phenomenon. Matthews stated that amongst a set of politicians competing for any office, a safe prediction is that the winner will be the one who most consistently presents a happy face, and an optimistic vision of the future characterized by growth of everything, free of any threats. The American electorate severely punishes politicians who attempt to present any unpleasant information. Presidents Nixon, Ford, and Carter all worked hard at trying to interest the electorate in a new national energy policy, to no avail. The four following presidents made no such effort (Reagan, Bush I, Clinton, Bush, II), and all were very popular, at least at times. This lack of interest in bad news runs all through American institutions and the electorate now. People who seek to be successful and advance rapidly in politics, corporations, or other bureaucracies discover that it pays to present a worldview free of problems and overlook or suppress information about impending threats. "Political knowledge"—that is, a view that will be appealing to some audience for political reasons—is frequently substituted for scientific knowledge. The capacity of society to make rational policy decisions is severely eroded when this political knowledge is treated as if it were scientific knowledge. The annual reports filed with O.P.E.C. on national oil reserves are examples. These statements have no geological or geophysical foundation, but are designed to improve the bargaining position on oil pumping rates with respect to other oil exporting nations. This bias in favor of the happy face is a particular problem for the systems community, because we tend to explore systemic properties of systems, rather than analyzing fragments of systems lifted out of context. Therefore, we discover effects resulting from interactions between variables of diverse kinds that would probably pass undetected by most people. The ultimate outcome is that we stumble into early warning evidence of impending threats which most of the electorate, politicians, and bureaucrats would rather not hear.

Highly sophisticated and meticulously planned psychological manipulation of the electorate is now a key feature of a winning strategy in elections. The United States population has now become filled with a vague dread and terror which affects behavior. This terror has been very carefully communicated to them, in association with the idea that one political party will be more capable of protecting the population than the other. A fact which has received little attention in the major media is that many of the leading administrators in the Department of Homeland Security have quit. This is odd, because being successful in attaining the stated goals of that department would be a marvelous stepping stone to the Presidency.

The dread expresses itself in the surprising number of people who wish to emigrate from the United States, and the increasingly widespread awareness that most other countries expect to be attacked by the United States and are therefore making ultra-high technology preparations. Also, about five times a week the media swamp us in data about wildly deviant behavior of a sort we never heard of prior to 1990. This uses up media time that otherwise might have been used to inform us about our national situation, something which the owners of the media obviously do not want to happen. A consequence is that when scholars present the public with results such as those in this chapter and this book, most people have great difficulty accepting reality for what it is, rather than paranoid fantasy.

2. A fragmented worldview, and fragmented institutions.

All our institutions, including our educational institutions, are organized to manage complex phenomena by fragmenting them into bits. For example, movement of people and goods into the United States is managed by the Justice Department (Immigration and Naturalization Service, the Border Patrol), the Treasury (The Customs Service), transportation oversight agencies (Federal Aviation Administration), the Defense Department, and the State Department (population, refugees, migration). An ironic feature of the system is that if you are invited into the United States by a major U.S. employer to take a leading high visibility job, you and your family have to jump through hoops of fire. Every member of the family requires an impeccable record with respect to crime, health, and mental health, and the principal wage earner may require a doctorate from one of the best universities in the world. If you immigrate into the U.S. to make a very large hole in Manhattan or the Pentagon, entering is simple and without any complications. Intelligence agencies are in the Departments of State, Treasury, Defense, Justice, and the C.I.A., yet no one in authority acknowledged the possibility of an event such as September 11, 2001, despite detailed warnings from the intelligence agencies of at least three other nations. Even internal alerts from two senior and well-regarded agents in the F.B.I. had no effect. Certainly, there was no attempt to coordinate information coming from multiple sources about the possibility of commercial jet airliners being flown into buildings, even when U.S. agencies were provided with lists of the intended target buildings. Hardly anyone is paid or promoted to examine how the constituent bits of systems work together. Since few people are taught how to do this, the process of bit-assembly goes undone in most situations.

Here is an example of what we discover when we explore causal pathways between types of variables not typically thought of as related. World population size is extremely well predicted by global crude oil production per person in the world per year, 28 years in the past. Fossil fuels are not only essential for manufacturing and powering all our vehicles, they also make possible the conversion of constantly increasing land areas to agricultural use, and increase the food productivity per acre by at least five-and-a-half times.

Since a number of entirely different techniques are available for predicting world crude oil production per person with very high accuracy, we can predict world population with high accuracy: oil production per person predicts 99.73 per cent of year to year variation in world population size, 28 years later. We discover that by 2085 the global human population will be about 1.5 billion, 23.3 per cent of the present level. I would not be prepared to bet that the huge drop in population will be accomplished peacefully. Americans are unaware of the population reduction experienced by other countries in previous major wars. To illustrate, in Russia in 1959, for 55 to 59 year olds, there were 2905 thousand men still alive, and 5793 thousand women, or 5793/2905 = 1.994 women per man. In East Germany in 1946, for 20–24 year olds, there were 707/309 = 2.29 women for every man still alive.

It seems worthwhile for the United States to take seriously the possible outcome of future wars. There is a feeling here that the United States should attack other countries to get their oil, as in the bumper sticker "Whip their ass and take their gas." Anyone who thinks that any other countries are not expecting to be attacked by the United States to get their resources needs to talk to more people from other countries. Only in the United States are there some people who assume we invaded Iraq for reasons other than to seize control over their oil. This issue is now turning up in the popular media of all countries. In *Foreign Policy* for May/June 2005, a letter to the editors concerning relations between China and the United States ends with this sentence: "Superpowers find it hard to cooperate when resources are scarce". *Walrus* (Canada) for April 2005 had this cover story: "As the Oil Runs Out, the U.S. and China Square Off in Alberta: What Card Will Ottawa Play?" My models show that in 2025, world oil production per person in the global population will be 47 per cent the 1980 level, so this issue may be worth some thought.

This is an appropriate point to summarize the most recent available statistics on the U.S. and world crude oil situations, as in the following text table.

Statistic	Year	United States Amount	Year	World Amount
Peak oil production	1970	3517 mill. bbl.	2000	24945 mill. bbl.
Most recent statistic	2002	2097 mill. bbl.	2002	24397 mill. bbl.
Peak oil production per person	1970	17.0 barrels	1973	5.44 barrels
Oil production per person Most recent statistic	2002	7.28 barrels	2005	4.49 barrels

Table 1. The current and future status of crude oil production in the United States and the world.

I remind readers who may be puzzled by these statistics that, as economists point out, what is important in a market is neither supply nor demand by themselves, but the relation between the two. Thus the key oil statistic is not production, but U.S. production per person in the U.S. and, by the same reasoning, global oil production per person in the global population. Thus, while global oil production increased to at least 2000, and perhaps to 2005 or 2006, global population is still constantly increasing, so global oil production per person worldwide has been decreasing since 1973. The United States population is protected from having to worry about the implications of this overall picture, because we receive a disproportionate share of the oil in the world.

The massive amounts of money being sent to other countries to pay for imported crude oil are paid for entirely by an explosive growth in outstanding consumer credit, which then translates into niggardly expenditures on education at all levels, and explosive growth in the costs of the criminal justice and incarceration systems, which have become winning competitors over the education system. A text table shows the astonishing consequences.

Year	Total number of prisoners incarcerated in all jails, prisons, and penitentiaries	Total number of degrees awarded, all levels and types	Prisoners per degree
1980	502 thousand	1731 thousand	.29
2002	2026 thousand	2491 thousand	.81

Table 2. In the United States, the number of incarcerated prisoners will soon exceed the number of degrees awarded each year.

This is a thought-provoking example of the type of phenomenon not widely publicized if no one is paid to study "the big picture." The problem is exacerbated because the outsourcing of technical jobs from the United States to other countries is discouraging students from taking scientific and engineering degrees. Even amongst business school M.B.A. graduates, there is rapidly growing interest in emigrating to Asia after graduation.[1] Young people are getting it figured out that "globalization" means that all good jobs are going to go to other countries so that there will be no future here. That, in turn, means that before long, U.S. national average consumer purchasing power will be too depressed for people to be able to buy whatever large transnational corporations are selling. A critic of that logic needs to come up with a completely convincing explanation for recent U.S. stock market behavior.

3. History has nothing to teach us.

This position seems to be a repeated feature of civilization noteworthy in Imperial Rome, Spain, and elsewhere that might be justified as wallowing in past glory. It leads to (a) a focus on the very recent past in any search for root causes of the current state of the societal system; (b) denial that there are feedback control mechanisms operating on multiple time scales, producing such phenomena as the multi-century rise and fall of a civilization, the 50–60 year wavelength Kondratieff economic/political/cultural "Long Waves," and a preoccupation with the brilliant decision-making by great leaders rather than natural forces as being the prime shaping force driving historical processes. Thus, a worldview more related to the Greek god Zeus than their Goddess Gaia is associated with the denial that history has anything to teach us.

Two examples for which there is widespread accurate documentation support a notion that is missing from much modern systems forecasting: there are very long lags in human affairs from cause to effect. A text table makes the case succinctly.

Incident	Year	Time lag (years)
Drake drills first oil well	1859	
Benz invents first gasoline car	1885	26
First patent for internal combustion car in U.S.	1895	36
4100 cars manufactured in U.S.	1900	41
Wright brothers first controlled and sustained airplane flight	1903	44
43 aircraft manufactured in U.S.	1913	54
268 airplanes operated in U.S. by scheduled air transportation companies	1928	69
Time lags in understanding that world oil production will peak around year 2000		
M. King Hubbert writes "2000" in widely available books	1962–1974	
"Peak oil" cited 3.28 million times by Google	March 22, 2005	
4.72 million times	April 16, 2005	
6.12 million times	May 14, 2005	

Table 3. The length of the time lag from the time a new idea or invention occurs to the time when the implications are widely understood and applied.

Thus, the fact that there is a 28 year lag from the year when global crude oil production per person in the worldwide population affects world population size should come as no surprise. There are very long lags from the time the world is in a certain state to the time that it is widely understood that the world is in that state. There were 6,500 Allied planes on runways ready to participate in mass attacks against Germany in November 1918. Yet the U.S. Dept. of Defense did not believe such an attack against Pearl Harbor could succeed prior to 7:49 A.M., Sunday December 7, 1941. An American, Billy Mitchell, had invented the idea of mass air attacks that would catch the enemy by surprise around 1917, about 24 years previous.

4. The shift from concrete random thinking to abstract sequential thinking.

David Hacket Fischer, in *The Great Wave*, Appendix O, noted how the social sciences have switched from dependence on empirical knowledge to knowledge generated by following logical branching trees.[2] This has fostered vast new literature on learning and thinking styles,[3] in which about thirty books by Edward de Bono are very helpful.[4] They discuss two very different research styles used to arrive at knowledge. In abstract sequential thinking, one idea follows from the previous idea, like the process of laying train track. The process gathers inertia as more track is laid, so the thinking process itself directs us to data we should be examining. In concrete random thinking, there is no track: we observe nature; data and experiments determine the paths along which our theories evolve. This latter trial and error type of thought and evolution of ideas turns out to be very powerful, because it leads us to contemplate phenomena and truths which our imaginations might not suggest. For example, extremely powerful computer algorithms for solving hyper-complex problems, as in finding the best equations and parameter values to describe long runs of historical data, are based on trial-and-error iterative procedures, not iterative procedures based on formal mathematics.[5]

5. The preoccupation with money, and overlooking non-monetary measures of societal functioning.

Variables measured in money cannot by themselves explain the dynamics of modern societies. Systems models that explain and predict all features of modern societies must also include variables that measure flows and storages of energy, matter, human populations, and other variables, if the models are to predict with high accuracy.

What Can the Systems Community Do?

To counter these problems eroding the capacity of society to forecast, plan, or manage itself, system scientists can bring new techniques of argumentation into the political arena.

1. We need to be surprising, and present new ways of looking at things that will catch the attention of mass publics. One way of doing this is to point out connections between system components that few other people have thought of, or to attract attention to anomalies when we compare phenomena from different fields, times, places, or different scales of time, space, or system organization. Simple, startling graphs are effective.

2. We need to present arguments in such a way that the data base is beyond reproach, and the technical means of manipulating the data is completely transparent and invulnerable to attack or criticism from scholars in any field.

Combating the "Happy Face"

For example, to combat the institutional and political inertia in dealing with the immigration problem, a useful approach is to compare the economic and demographic situations of the United States, and on the other hand, the whole world, and immigration-restrictive, economically successful countries such as China. The difference between the demographic situation of the United States and the rest of the world is startling, once explored. To begin with, we need a model for forecasting population with very high accuracy. Colin Campbell and I developed such a model through exhaustive computer analyses of the post-1599 global population estimates. To illustrate the utility of this model, in Table 4 values computed from the model are compared with the most recent large table of world population estimates. Considering just estimates of the global 2000 population, they were 6241 million in Table 1402 of the 1989 Statistical Abstracts of the U.S., and 6080 in Table 1324 of the 2001 edition. Our model shows world population peaking by 2008 and dropping rapidly thereafter. To compare this with the U.S. situation, we take a 13-year moving average of U.S. annual population growth rates, to smooth out year to year fluctuations and clarify the long run pattern. The U.S. population growth rate stopped declining about 1987. Clearly, because of the high level of immigration, it is unlikely that the United States population will decline any time in the

Year	Population estimate (Millions)	Value computed from model (Millions)
1950	2516	2565
1955	2752	2779
1960	3049	3031
1965	3358	3325
1970	3678	3661
1975	4033	4036
1980	4457	4441
1985	4855	4865
1990	5284	5291
1995	5691	5695
1998	5925	5918
1999	6003	5987

Table 4. Comparison of recent global population estimates with values computed from the Campbell-Watt model which describes period 1650 to present with very high accuracy (unexplained variation is 5 parts in 10,000). The population estimates are from Table 1350 in the Statistical Abstract of the United States for 2000, and comparable tables in other editions, and are based on the International Data Base of the U.S. Bureau of the Census.

foreseeable future, while global population is projected to be declining rapidly in just 30 years. The difference is because the rate of growth of the U.S. population can not drop, no matter how low the drop in fertility rate of women born here; it is being more than compensated for by a tidal wave of immigration from all over the world. Thus, while the world population will be shrinking in the last half of the 21st century, the U.S. population will be growing explosively. By 2075 it will be about 555 million, more than twice the present population. The rest of the world will be getting richer, and the United States will be getting poorer, because of the well-known inverse relation between rate of population growth and rate of growth in affluence per person.

Afghanistan from 1990–2000 had the highest rate of growth in population of any nation, 5.6 per cent per year. This enormous population growth rate will seem outlandish to some readers, and some statistical compendia give far lower estimates. The 5.6 per cent comes from the International Data Base of the U.S. Census

Bureau. Two other sets of statistics from the same data source may make this 5.6 per cent more credible. First, extremely high growth rates are characteristic of states known for a high level of violence. The average annual rate of growth between 1900 and 2000 was 5.7 per cent for the Gaza Strip, 4.8 per cent in the West Bank, and 3.7 per cent in Yemen. Second, readers will have noticed in television news from violent places, there are a great many young men. In the United States in 2000, 21.2 per cent of the population was under 15 years old. For Afghanistan in 2000 it was 42.4 per cent, and in 2010 it will be 53.0 per cent. Indeed, one of the great systems thinkers of all time, H. G. Wells, pointed out the relation between societies with an unusually large numbers of young men, and subsequent high probability of war. This is a principal motive for arguing for limitation of population growth rates. High population growth rates produce a higher proportion of very young people than their elders can readily educate in traditional cultural values.

Television news shows the awesome impoverishment of the Afghanistan landscape. The last trees will be gone in about 12 months, and trees are a major source of fuel for the nation, fossil fuel being too expensive.

A Fragmented Worldview

A fascinating example of the consequences of a fragmented worldview and fragmented institutions concerns the fate of education in the United States. We think we are a New Economy/High Technology Society; as such, you would expect us to have an educational system that would prepare people for work in such a society in numbers adequate for our needs. The picture becomes more startling the more you dig into it. The number of people awarded all Bachelor's, Master's and Doctor's degrees in the United States grew by 61 per cent from 1970 to 1999; the number incarcerated in all jails, prisons and penitentiaries grew by a factor of six. All kinds of other social statistics reflect this competition between two major industries, one of which is the clear winner. For example, there is enormous growth in the number of attorneys, judges, police, guards, wardens, detectives, and sheriffs in contradistinction to much slower growth in numbers of professors, doctors, and teachers. The number of Bachelor's degrees awarded in education in 1999 was 60 per cent what it was in 1971. A similar picture emerges when one explores the trends in degrees awarded in fact-based fields as opposed to belief-based fields.

Other interesting phenomena are exposed when we take a systems approach to debt. What causes debt, and what causal pathways does it affect? There are some curious omissions in any discussion of the present state of the United States. One is the enormous amount of money spent on imports, net of exports, of mineral fuels: 1.3 trillion dollars after 1969. Who paid for this, and how? The answer is that it was paid for by consumer debt, which increased by 1.3 trillion dollars after 1969. Now what is the effect of that debt? That takes us to the next section, on what we can learn from history.

History Does Have Much to Teach Us

Debt is an orphan topic amongst politicians and the media: no one seems to want to talk about it, even though enormously large amounts of money are involved. We keep hearing about how "The Fed is our friend" and "You can not fight the Fed". In other words, any time it wishes, the Board of Governors of the Federal Reserve Board, chaired by Alan Greenspan, can stimulate the economy by decreasing the interest rate, so as to increase the amount that people will borrow. No mention is made of conditions under which this would not be true. In fact, once any individual or institution has borrowed against all the collateral they have ever had or will have, they can borrow no more, no matter how low the interest rate. History can teach us about this. From 1920 to 1929, there was an enormous increase in borrowing, as there has been since 1970. When the Great Depression began, the monetary authorities began lowering the interest rate, and kept at it until 1935. As we all know, this did not prevent the Great Depression. All the talk about the help from The Fed is one more example of putting on the Happy Face, or of political knowledge brushing aside scientific knowledge.

Now we can put together a story about debt. The importation of vast quantities of crude oil to replace domestic reserves, which have been declining since 1970, created a huge loss to the economy; a subtraction from, not an addition to, Gross National Product. This was paid for by an increase in consumer debt. Because of this debt, which has now reached a limit set by collateral, no action by the Federal Reserve Board can stimulate the economy, contrary to what we are told by the interests that make their living off commissions from selling shares in corporations, which only rise when sales grow.

Note that this argument depends on linking domestic geological reserves of crude oil with energy imports, the cost of those imports, the debt created by that cost, and the effect of that debt on economic growth once it became large enough to restrict further growth of borrowed money. This argument is essentially interdisciplinary, not intra-disciplinary, and hence is the appropriate turf or territory to be occupied by systems scientists.

By examining history, we learn that all systems phenomena fluctuate on several different time scales. We illustrate using the United States Wholesale (Producer) Price Index (1967 = 100.0). If we take 39-year moving averages of this statistic, then find the mathematical equation that describes that trend with higher accuracy than any other equation and project the trend, we get the graph in Figure 1. One reviewer found this graph incredible because of the sharp upward move in prices after 1972. To make this more credible, the graph can be compared with Table 797 in the 1984 Statistical Abstract of the United States. This table shows that consumer prices never increased more than 5.9 percent per year from 1960 to 1972. Here are the percent per annum increases in consumer prices for the five most startling years from then to 1982.

Year	Percent per year increase in consumer prices
1974	11.0
1975	9.1
1979	11.3
1980	13.5
1981	10.4
1982	

Table 5. An example of the violent year to year oscillations in the rate of growth in consumer prices.

This huge increase in prices was made possible by a historically unprecedented increase in the M2 measure of the money stock from 1974 to 1983, perhaps to ameliorate the effects of a huge increase in crude oil prices brought on by the O.P.E.C. embargoes.

Fluctuation can occur on the longest time scale, analogous to the rise and fall of the Roman Empire. If actual historical W.P.I. data are divided by that long run trajectory, we see a second pattern of fluctuation with a wavelength of 30 to 60 years (Figure 6). This graph certainly challenges the bombastic battle cry, "We have conquered the business cycle." Now a long downturn from 1990 to 2003 or later

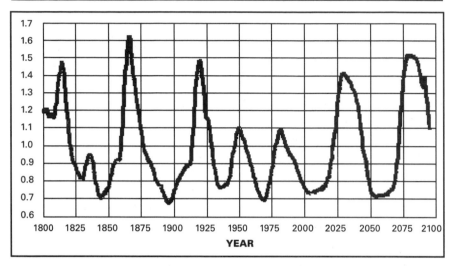

Figure 1. Past and projected seven-year moving averages of U.S. Wholesale Price Indices, divided by their long run trajectory. Past moving averages were computed from U.S. Department of Commerce data series E-23, and comparable data in recent government documents. Future indices were generated using the simulation model discussed in the text.

becomes entirely expectable. This pattern of variation can be decomposed into a number of components. For example, there is a feedback control loop with a time delay of 22 years, such that 22 years after a peak, there is always a very low value, and the highest peaks always occur 22 years after a very low value. This 22 year time delay in economic affairs occurs because 22 is the age at which women have the largest number of live births per thousand women per year. Thus, 22 years after a large number of babies have been born, there will be an unusual number of women seeking upward economic mobility in the labor force, competition amongst them will be intense, and that will drive down wages and prices. That will depress birth numbers, so that 22 years after that, there will be few 22 year old women, and they will have a large number of babies per thousand women per year, reinitiating the cycle. The scatter of the points in graphs of the raw data tells us that feedback control is occurring on other time scales.

We can build a mathematical model of as little or as great complexity as we choose to predict the economy. Figure 1 shows the result when only three variables were used to make the prediction. Note that even at this crude level, future fluctuations appear to be qualitatively similar to the fluctuations of the past. It is noteworthy that this type of model, using the concept of variation on multiple time scales, does not predict the immediate economic upturn we kept hearing about during 2000 and 2001 from the people who earn a living by selling shares of stock.

Styles of Learning and Thinking

Of immense significance for systems scientists is the fact that forecasting models in almost all areas are a disaster, a subject developed in a large critical literature. An explanation is that forecasting models are not based on an empirical analysis of how the world works, but rather on our beliefs about how the world works. One of our hidden, unexamined assumptions about systems behavior is that causal pathways do not flow between variables of different types. That is, economic variables would not be affected by demographic variables, and vice versa. Are depressions accidents? In the past, it has taken about 29–32 years from the time a boy was born until he had made enough money to satisfy the parents of a young lady he wanted to marry. It is of interest that the depression of 1843, which affected many nations, fell 29 years after the end of the Napoleonic wars, which was followed by a large surge in births. The depression of 1896 fell 32 years after 1864, which was followed by another surge in births. That is, the surges in births which follow all wars reset the timing mechanism for Kondratieff Waves.

The point here is that if we are prepared to let our thinking follow the data from cause to effect repeatedly down a causal chain, without regard to the discipline that would claim each variable as its own, we can understand, forecast, plan, and manage systems.

Preoccupation with Money and Monetary Measures of Society

Another question was unanswered. Why did the wavelengths vary in Figure 6 ? The length of a downturn decreases the faster the rate of growth in energy production per person per year. The faster that rate grows, the speedier will be the growth in inventories of everything from new houses to cars, television sets, or clothes dryers. The faster the growth in inventories, the sooner economy-wide market super-saturation and depression will arrive. In short, economic waves shorten when there is rapid growth in energy production, all other things being equal, and lengthen when growth in energy use per person is very slow. The days of rapid growth in energy use and production are over, so we will now be going back to the long wavelengths of the 18th and 19th centuries.

Overview

It is not enough that system scientists develop new ways of looking at things. We must also become more proficient at explaining ourselves to electorates, politicians, and corporate executives. I find that it is necessary to put about a hundred times the effort into refining the technique of explanation and argumentation as is required to discover and develop an accurate mathematical model for a phenomenon. Through trial and error, I have found that a useful approach is to hire as editors brilliant students from areas remote to those of the authors being edited. We also need to become very active in the political arena, and in communicating with the media.

Notes

1 White, E. 2005. For M.B.A. students, a good career move means a job in Asia. *Wall Street Journal.* May 10; p. B1.
2 Fischer, D.H. 1996. *The Great Wave. Price Revolutions and the Rhythm of History.* New York: Oxford University Press.
3 Dryden, G. & J. Vos. 1994. *The Learning Revolution.* Rolling Hills Estates, California: Jalmar Press.
4 De Bono, E. 1977. *Lateral Thinking. A Textbook of Creativity.* New York: Penguin Books.
5 Schwefel, H.P. 1994. *Evolution and Optimum Seeking.* New York: Wiley.
6 USBC. 2000. *Statistical Abstract of the United States 2000.* Vol. 200th ed. Washington, D.C.: U.S. Bureau of the Census, U.S. Govt. Printing Office.

12

Economics Weak and Strong

Ecological Economics and Human Survival

Andy Bahn & John Gowdy

Introduction: Humans and the Rest of the World

The recent Global Environmental Outlook (GEO-3) report on environmental trends since 1972 by the United Nations forecasts dire consequences from continued population and economic growth around the world over the next thirty years. The predictions are familiar and include the rapid expansion of urban areas, forest destruction, soil degradation, depletion of freshwater sources, and the increased burning of fossil fuels. These assaults on the environment will lead to major losses in biodiversity, famine, food and water shortages, and global climate change due to the buildup of greenhouse gases.[1] These trends have already damaged local and global ecosystems to the extent that essential life-support systems are threatened.

GEO-3 reports four scenarios exploring what the future could look like in thirty years based on different global policies toward the environment. The scenarios span developments in many areas including population, economics, technology, and governance. The four scenarios—termed Market First, Security First, Policy First, and Sustainability First, respectively—all forecast continuing negative trends over the next two decades. The report is unusually pessimistic in that it concludes that the benefits of even a "sustainability first" policy approach will not occur for decades because most of the negative environmental changes expected over the next thirty years have already been set in motion. Even if environmentally friendly policies were put into effect today, CO_2 concentrations will continue to increase until 2050, freshwater shortages will continue, and coastal pollution will increase. The consequences of a "markets first" approach are far more negative and include increases in poverty and inequality, a rising flood of immigrants, and virtually uncontrolled exploitation of the world's ocean and forest ecosystems.[2]

Given the central role of the market economy in dramatically changing the ecological and social conditions on planet Earth, it is natural that the field of economics should be a leading player in Survival Research. The growth of the market economy, including increased consumption rates, rapid urbanization, and growing income disparities within and between nations, lies behind all the negative environmental trends of the past thirty years. All these important trends, then, fall within the domain of economics and should be the focus of intensive efforts by economists to understand the human impact on the natural world. But over the past hundred years or so, the subject matter of economics has steadily narrowed from "a study of man's action in the ordinary business of life" (Alfred Marshall) to the study of "human behavior as a relationship between ends and scarce means which have alternative uses" (Lionel Robbins). Neoclassical economics is a useful description of how markets allocate scarce resources, but it is of little use when it comes to policy recommendations for human survival. Contemporary economic theory is both a description of how stylized markets allocate resources, and an ideological justification for the superiority of the unfettered market economy. Survivability research requires both an understanding of how market economies operate, and an understanding of how the force of market exchange might be diverted from its current destructive path to one of environmental and social sustainability. The new field of ecological economics explicitly addresses the survivability question.

This paper is organized into three parts. First, we describe the historical development of the human economy from the Paleolithic age through the Neolithic revolution and up to the modern industrial age. Some of the current negative trends described by GEO-3 have their origins in pre-industrial societies, and there is much to be gained from understanding these patterns. Second, we describe the neoclassical model of market exchange (called "welfare economics"), and the virtues and shortcomings of that model. In this section we contrast the neoclassical definition of "weak sustainability" with the ecological economics alternative of "strong sustainability." We conclude with a discussion of the prospects for human survival and suggest some policies promoting sustainability of the Earth's life-support systems.

Technology and Human Survival

Some of the same forces that enabled our Paleolithic ancestors to survive dramatic environmental changes—the ability to substitute one resource for another in the face of increasing scarcity, and the extensive use of technology—today threaten the life support systems of the planet. Later in time, Neolithic societies were able to escape the confines of local ecosystems, but they were characterized by a pattern of overshoot and collapse. Today, as industrial civilization covers the entire globe, we are once again confronted by environmental limits—this time, the limits of the life support systems of planet Earth.[3] The history and prehistory of past human societies offer important lessons for the current state of the biosphere and the global economy.

Our earliest human ancestor is *Homo erectus,* who appeared in east Africa about 1.9 million years ago. *Homo erectus* was about as tall as modern humans, had a brain almost as large as ours, used fire, and possessed a recognizable stone tool technology that lasted well over one million years. It is now widely accepted that *Homo erectus* evolved a large brain and a sophisticated technology in a rapid burst of evolution in response to changing climatic conditions in Africa. With fire and ingenuity, *Homo erectus* spread relatively rapidly throughout Europe and Asia. Some evidence exists that this human species survived in Southeast Asia until about 30,000 years ago. The emergence of fully modern humans, or *Homo sapiens,* is relatively recent, perhaps as early as 100,000 years ago. The rapid transition from *Homo erectus* to the larger brained *Homo sapiens* with a much more sophisticated stone technology, was once again most likely due to rapid climate change in Africa. Our early *Homo sapiens* ancestors were apparently very similar to us in terms of the capacity for abstract thought and sophisticated language skills, and they had all the required traits to enable them to eventually develop the societies we see around us today. It is conceivable that if we could go back in time 100,000 years and bring back an infant from one of our ancestors, the infant could be raised in our modern society to be a functioning member of the industrial era. Likewise, we could leave one of our infants to be raised within the nomadic tribe of hunter and gatherers, and she would perform equally well within that ancestral society.

Homo sapiens lived as nomadic hunters and gatherers for the majority of the time our species has inhabited planet Earth. Although evidence of the systematic control of wild plants and animals goes back at least 20,000 years, agriculture as the basis of economic and social organization does not occur in the fossil record until around 10,000 years ago, when it appeared simultaneously in different regions around the world. The exact cause of the beginnings of agriculture may never be known. Mark Cohen argues that, from the beginning, nomadic groups of hunters and gatherers put increasing stress upon their environmental resources whenever they first moved into a new region. They adapted to increasing scarcity by migrating to new unoccupied regions or by shifting the source of their caloric intake from a vanishing, relatively easy source of calories to food sources requiring a greater effort to capture and harvest. A likely catalyst for the switch from hunting and gathering to agriculture is, once again, rapid climate change. The sudden and simultaneous appearance of agriculture in different regions of the world points toward some exogenous worldwide shock. And once more, *Homo sapiens* adapted to an inability to support existing populations using the current subsistent mode of hunting and gathering by intensifying technology.[4]

The distinguishing characteristic of the agricultural "revolution" is the shift from living off the seasonal flows of calories that described the hunter and gatherer lifestyle to living off of stocks built up and stored for use during the off-season. Hunter-gatherers were nomadic and followed seasonal migrations of animals and ripening wild fruits, grains, and vegetables. They lived off of the flows that the land offered them. This was a sustainable lifestyle, as long as their populations remained stable. Once human groups began living off of stocks, sustainability became problematic. To increase existing stocks, larger populations were needed to work the land. And as populations grew, stocks had to be increased to support them. This positive feedback loop did not describe an equilibrium state, but rather one characterized by increasing returns to scale.

An advantage of the new sedentary lifestyle was the ability to develop tools that did not have to be hauled from camp to camp during regular migrations. This sedentary existence allowed the development of material "capital" to extract resources from the land through hunting and agriculture. Greater populations could be supported and more hands were available to work the land.

The larger group size in early agricultural societies also led to social stratification. The existence of a surplus meant that particular individuals or groups of individuals could control that surplus. Individuals could also specialize in certain tasks resulting in greater returns to the group. Finally, as groups created surpluses but lacked resources that were not in their immediate surroundings, the climate for trade developed. Trade networks were formed between agricultural regions for resources, technology, and ideas. As trade flourished, there were even greater returns to investments in material capital. There was an expanding loop that encompassed larger and larger geographical areas. Populations grew, which created an environment for more specialization; technology and capital stocks increased, meaning more calories could be extracted per hectare, which fed the growing population; and so on.

The use of increasingly complicated technology to overcome resource scarcity led many of these early civilizations to overshoot and collapse. These include such diverse cultures as early Sumer and Mesopotamia, New World civilizations such as the Maya, and isolated Pacific Island cultures such as Easter Island. All these cultures followed the same pattern of colonization, rapid growth of population, soil depletion, and eventual collapse and rapid population decline.[5] Populations were either devastated by famine or migrated into surrounding areas, bringing their technologies with them to begin anew. The race between technology and scarcity that intensified with agriculture continues up to the present day, as greater populations need to be supported and new technologies need to be developed to support them.

The shift to fossil fuels is another example of a major advancement in the technological history of *Homo sapiens*. They provided a substitute for the over-harvesting of many renewable resources, such as timber for firewood. It also increased the ability to perform work on the land and, thus, the ability of the human economy to support a larger population. However, fossil fuels are non-renewable and their use has caused the global changes in climate we are currently experiencing, and which ominously face us over the next few generations as greenhouse gases accumulate in our atmosphere at an increasing rate.

The use of fossil fuels made possible the industrial age and the incredible growth of material output during the past 150 years or so. A direct result was the emergence of the market as the dominant form of economic and social organization. The market system has its roots in the agricultural revolution and the early trade patterns established by agricultural societies. But in the past two hundred years the market economy has evolved into something very different than any past human culture experienced. Understanding how markets work and how the market economy has changed the relationship between humans and the natural world is essential to the survival of our species.

Sustainability Weak and Strong

The dominant school of economics in North America and Europe is called neoclassical economics. It offers a stylized description of the allocation of goods and productive inputs in a market economy based on the notion of "Pareto optimality". Given a fixed amount of goods to be allocated and a particular initial distribution of these goods, free exchange will result in a situation in which no one can be made better off without someone else being made worse off. In economic jargon, no further trading can increase the utility of one person without reducing the utility of another. This end result of free market exchange is Pareto optimality. The analysis can be easily extended to include the production side of the economy—firms exchange inputs to maximize output to the point where no further exchange can increase the output of one firm without reducing the output of another. Two critical points about neoclassical welfare economics should be kept in mind. First, the theory is useful as a description of impersonal market exchange. Secondly, welfare economics also shows the folly of relying on markets to insure social and environmental sustainability. The neoclassical model of welfare economics—mimicking the real market economy—assumes, among other things, that individual decisions are independent of their social and environmental context, the future is worth less than the present, and everything is substitutable. Although the welfare economics framework has severe limitations in describing many features of the contemporary economy, it shows the way impersonal market exchange works, and it shows why the natural world is being reduced, piece by piece, to fodder for the industrial economy.

The problem with neoclassical welfare economics is that economists use the model not as a positive description of markets, but as a normative prescription of how the world should be. Instead of addressing the innate conflict between markets and the long term survivability of the human species—clearly shown by welfare economics—economists instead advocate correcting "market failures" to achieve Pareto optimality. Efficiency in allocation is assumed by neoclassical economists to be the only "scientific," or "positive," policy goal. But as Bromley so eloquently argues, the goal of efficiency is ideological and therefore is "normative" rather than "positive," as most economists contend.[6]

The mainstream economics definition of sustainability is called weak sustainability. According to Brekke, "A development is said to be weakly sustainable if the development is non-diminishing from generation to generation."[7] To make this definition operational, neoclassical economists narrow the determinants of sustainability to those that contribute to sustaining the output of the market economy, focusing on those that are important to intergenerational equity. To determine the value of social welfare, economists equate welfare with aggregate consumption measured by a nation's Net National Product (NNP), defined as Gross National Product (GNP) minus depreciation. The formula for determining the sustainability of a country's economy is suggested by Pearce and Atkinson:

$$Z = S/Y \text{-} d_m/Y \text{-} d_n/Y$$

where Z is an index of sustainability, Y is GNP, S is (national) savings, d_m is the total depreciation of man-made capital, and d_n is the total depreciation of natural capital. An economy is determined to be weakly sustainable if $Z > 0$.[8] Rearranging the above equation, we can also define an economy as weakly sustainable if:

$$S > d_m + d_n$$

Since total savings in an economy is equal to total investment, and increases in welfare (consumption) can only be obtained by increases in NNP, investment in human or manufactured capital are the only means to increase aggregate consumption. The depreciation of the stock of non-renewable resources is not recoverable. Thus, if total savings are invested in human and manufactured capital at a rate equal to or greater than the depreciation of natural capital, the total portfolio of the capital stock will be non-declining.

This model implicitly assumes manufactured capital is a perfect substitute for natural capital, since savings can only be used to increase the stock of man-made capital.[9]

The framework of welfare economics is used by prominent economists to make policy pronouncements about some of the most serious environmental problems we face. In the case of global warming, for example, economists assume that a stable climate is just another commodity that gives people "utility"—in the same category as a new SUV or a European vacation. In this framework, the economic problem is choosing the utility maximizing mix of current consumption and investment in future climate stability. Using the welfare economics approach, it is no wonder that William Nordhaus concludes that it is too expensive to take even moderate measures to slow global climate change: "A vague premonition of some potential disaster is insufficient grounds to plunge the world into depression."[10] Neoclassical economists are equally dismissive of the problem of biodiversity loss. As Nobel laureate Robert Solow puts it, "History tells us an important fact, namely, that goods and services can be substituted for one another. If you don't eat one species of fish, you can eat another species of fish."[11] Biological species, like climate stability, are just commodities to be used and discarded as the market dictates.

The danger of relying on markets to dictate resource use is illustrated by the example of the South Pacific Island of Nauru. The case of Nauru shows that the standard economic definition can determine a country to be "sustainable" even if it completely depletes its entire stock of natural capital, as long as it invests at a rate equal to the rate of loss of these resources. At the turn of the last century it was discovered that this small island (with a circumference of 20 kilometers) contained some of the world's richest phosphate deposits. Over the last hundred years the island has been almost totally mined out, leaving an uninhabitable interior of jagged pinnacles and deep pits. The once beautiful island is environmentally devastated, with a catastrophic loss of biodiversity, exhaustion of its fresh water, and soaring interior temperatures. In return for selling off their island home, the people of Nauru received a substantial amount of money that was deposited in a trust fund once estimated to be worth over $1 billion. According to the formula for weak sustainability, Nauru was a shining example of a sustainable economy. Its savings rate was much higher than the rate of depletion of its natural capital stock. Unfortunately, some bad investments and the Asian financial crisis wiped out the trust fund and left the islanders with no money, a ruined environment, and a very uncertain future.[12]

A field challenging the dominance of neoclassical welfare theory is the emerging discipline of ecological economics. Ecological economics moves beyond the narrow framework of market exchange to examine economic activity in its larger social and biophysical context. Without minimizing the formidable task we face in stabilizing the planet's life support systems, we believe that the relatively new field of ecological economics can provide policy guidance. The alternative to the weak sustainability prescription of standard economics is strong sustainability.[13,14] Strong sustainability recognizes that human decision-making takes place within particular social systems, and that human beings are biological species with certain basic physical requirements for existence. For most people these statements are self-evident. Standard welfare economics, however, either denies them or considers them to be irrelevant. In spite of valiant attempts to sharpen strong sustainability, it is impossible to conceptualize the concept within a mathematical model or even in a clearly defined set of rules for sustainability. As Beckerman correctly points out, defining strong sustainability is impossible when analyzed through neoclassical models. Strong sustainability should be regarded as a goal or a process towards which the world strives, rather than a formal framework from which we immediately implement global environmental policies.[15]

Strong sustainability recognizes that we are in an era of "post normal" science.[16] Traditional science is needed more than ever to guide policy-making. But decisions now are characterized by a high degree of uncertainty, limited information, and a very short time frame within which to operate. Dealing with the problems of survival science necessitates stepping outside the framework of efficiency, marginal change, certainty, and tradeoffs. Instead we should encourage policies promoting conservation, decreased consumption rates, substitution of renewable for non-renewable resources, and stable (or zero) growth. Measures promoting the preservation of natural stocks while substituting the use of renewable resources or flows should be implemented whenever feasible. In this spirit, let us revisit the issues of climate change and biodiversity loss.

As the Nordhaus quote above illustrates, conventional economic analysis tells us that it is too expensive to stabilize the climate. But the climate scientist Stephen Schneider and the economist Christian Azar revisited Nordhaus' analysis and came to a quite different conclusion.[17] Using Nordhaus' own assumptions, they found that the cost of stabilizing year 2100 emissions at twice the pre-industrial concentrations—between $1 trillion and $8 trillion—would be negligible compared to the predicted growth of economic output. Without curbing emissions at all, economists predict that the world as a whole will be ten times richer by the year 2100. Adding the costs of climate stabilization will postpone this ten-fold increase by only two years; 2102 rather than 2100. The pronouncements of mainstream economists on the global warming issue are driven more by neo-liberal ideology than by scientific facts and good judgment.

Biodiversity loss is the other long-term threat to human survival. Humans evolved in the web of biodiversity on planet Earth, and current human activity is shredding this fabric of life. Before humans appeared on the planet, the rate of species extinction was about one per million per year, about the same rate as new species came into existence. Today, human activity has upped the rate of extinction to about 1000 per million per year. Most of this extinction is occurring in a few biodiversity hotspots in tropical rainforests and coastal areas. Compared to the cost of stabilizing greenhouse gas emissions, the estimated cost of protecting the world's biodiversity hotspots is miniscule, about $28 billion. Twenty-five of these hotspots have been identified as critical, and these cover only about 1.4 percent of the planet's land surface. But if we save that 1.4 percent, we can save a large portion of species listed as endangered.[18]

These examples show that taking major steps toward our future survival would not "plunge the world into deep depression." In view of the importance of climate change and biodiversity loss to our very existence as a species, it seems incredible that we cannot take the necessary steps to address these problems. The fact that world leaders hold the environment in such low priority shows the power of the ideology of the global market economy—an ideology glorifying consumption, production, and economic wealth over long term survivability.

Conclusion

As the review of our evolutionary history showed, two things stand out. The first is the importance of climate in our physical and social evolution. The second is the striking characteristic of our species to adapt to changing conditions through technological and social innovations. If we can shake ourselves out of the ideological straightjacket of conventional economic theory, there is hope that we can survive this century with our environment and our social fabric intact.

If we step out of the mindset of welfare economics, what would our reference of judgment be? A good starting point would be Georgescu-Roegen's admonition "love thy species as thyself."[19] According to Georgescu-Roegen, the goal of economic life should be to minimize regrets, not maximize utility. In terms of policy, a good framework is Daniel Bromley's notion of "regency." Bromley points out that the present generation stands as dictator over those living in the future. Just as in the case of regents protecting a kingdom until a young monarch comes of age, it is the duty of people living today to leave future generations the means to insure their well-being.[20] We may not know the preferences of people living in the future, but we do know something about their physical and emotional requirements as members of the human species. In this framework, the question becomes not "how much to leave," but rather "what to leave." The idea of regency can also be extended to the planet's other species and ecosystems through the notion of "stewardship". Stewardship recognizes that humans have a responsibility to the rest of the natural world. According to University of Maryland professor of public policy Peter G. Brown:

"The task before us is monumental: to re-envision our place in the world, from lords and masters to citizens and stewards. Nevertheless, the building blocks are within our reach. Our forebearers have given us the tools to refashion our future: the rule of law, science, the market, private property, a free press, the protection of the weak. These and a myriad of other hallmarks of our progress and vision as a species are essential elements in a refashioned future in the commonwealth of life. With courage and resolve we can reground education, reinvent industrial society, re-design economics, rediscover trustee government, and redirect civil society in service to the commonwealth of life." [21]

While economics has maintained its claim as "the queen of the social sciences," and continues to have a tremendous influence on business and governments around the world, the cracks in the foundations of its welfare core theory are threatening to topple the entire neoclassical paradigm. For over one hundred years economists and others have questioned the tenets of welfare economics with little effect. Today, however, criticisms of the standard model spring more and more from the pens of orthodox economists. Ecological economics, by placing the study of economics squarely within human society and ecosystems, can lead the way to make economics both scientifically credible and politically relevant.

Notes

1 United Nations Environmental Program. 2002. *Global Environmental Outlook* 3. London: Earthscan.

2 *Ibid.*

3 Eldredge, N. 1995. *Dominion.* Berkeley: University of California Press.

4 Cohen, M. 1977. *The Food Crisis in Prehistory.* New Haven: Yale University Press.

5 Ponting, C. 1993. *A Green History of The World.* London: Penguin Books.

6 Bromley, D. 1990. The ideology of efficiency: Searching for a theory of policy analysis. *Journal of Environmental Economics and Management* 19(10): 86–107.

7 Brekke, K. A. 1997. *Economic Growth and the Environment: On the Measurement of Income and Welfare.* Cheltanham: Edward Elgar.

8 Pearce, D., & G. Atkinson. 1993. Capital theory and the measurement of sustainable development: An indicator of "weak" sustainability. *Ecological Economics* 8 (2): 103–108.

9 Ayres, R., J. C. J. M. van den Bergh, & J. Gowdy. 2001. Strong versus weak sustainability: economics, natural sciences, and 'consilience'. *Environmental Ethics* 23: 155–68.

10 Nordhaus, W. Two years to save the world. *New Scientist Online.* June 12, 2002.

11 Solow, R. 1993. Sustainability: an economist's perspective. In: *Economics of The Environment.* R. Dorfman & N. Dorfman, eds. New York: Norton.

12 McDaniel, C. N. & J. Gowdy. 2000. *Paradise for Sale: A Parable of Nature.* Berkeley: University of California Press.

13 Daly, H.E. 1995. On Wilfred Beckerman's critique of sustainable development. *Environmental Values* 4(1): 49–55.

14 Gowdy, J. 2000. Terms and concepts in ecological economics. *Wildlife Society Bulletin* 28(10): 26–33.

15 Beckerman, W. 1994. "Sustainable development": is it a useful concept? *Environmental Values* 3(3): 191–209.

16 Funtowicz, S. & J. R. Ravetz. 1993. Science for the post normal age. *Futures* 5(70): 739–755.

17 Azar, C. & S. Schneider. Are the economic costs of stabilizing the atmosphere prohibitive? *Ecological Economics* 42: 73–80.

18 Wilson, E.O. 2002. *The Future of Life.* New York: Alfred Knopf.
19 Georgescu-Roegen, N. 1976. *Energy and Economic Myths.* New York: Pergamon Press.
20 Bromley, D. 1998. Searching for sustainability: the poverty of spontaneous order. *Ecological Economics* 24: 231–240.
21 Brown, P. 2001. *The Commonwealth of Life: A Treatise on Stewardship Economics.* Montreal: Black Rose Books.

13

Historical Perspectives on Global Ecology

J.R. McNeill

> *It happens then as it does to physicians in the treatment of consumption, which in the commencement is easy to cure and difficult to understand; but when it has neither been discovered in time nor treated upon a proper principle, it becomes easy to understand and difficult to cure. The same thing happens in state affairs; by foreseeing them at a distance, which is only done by men of talents, the evils which might arise from them are soon cured; but when, from want of foresight, they are suffered to increase to such a height that they are perceptible to everyone, there is no longer any remedy.*
>
> —Machiavelli, *The Prince* (1513)

I. Introduction

Machiavelli's wisdom concerning affairs of state applies with equal or greater force to affairs of global ecology. While problems remain easy to solve, they are difficult to recognize; and by the time they are obvious, most of them are exceedingly difficult to solve. One way to make this unhappy problem slightly more tractable is to take history seriously into account in matters of ecology. This article will attempt that, by taking the measure of environmental change in the last century or so and by casting light on the various relationships between environmental shifts and social, economic, intellectual, and political forces, with a decided emphasis on the last two.

Students of contemporary ecological issues are sometimes alert to the existence of history, but they often use the past simply as a storehouse of cautionary tales. The three most common such tales are probably those of the Indus Valley civilization, the Mayans, and the Greenland Norse. The first two were irrigation-based agricultural civilizations which suddenly vanished—or at least urban life, complex and hierarchical social structure, irrigation, literacy, and high culture vanished. In the case of the Indus, the collapse occurred by 1800 B.C.; the Maya collapse—if one can use that term to describe a decline that spanned more than a century—is usually dated to the 9th and 10th centuries A.D. The environmental interpretation of the Indus and Maya collapses emphasizes the vulnerability of irrigation-based societies. The Indus irrigation system may have suffered from salt buildup (as irrigation systems normally do), or from climate change or catastrophic floods. The Mayan agricultural system, based on so-called raised fields, may have proved unable to cope with prolonged drought. The Greenland Norse were a small-scale society, a Viking outpost that went extinct in the early 15th century. The Norse, or so the argument goes, failed to adapt to climate change, in particular the cooler temperature of the Little Ice Age (c. 1300–1850) that made their cattle and farming economy increasingly unrewarding.

Each of these interpretations might be right. But the problem is we do not know and are unlikely ever to know.[1,2] The written record is slender in each case, and archeological evidence alone cannot plausibly provide definitive answers. In general, the tendency to award ecological change a prominent place in explaining historical change, especially declines and falls, is inversely proportional to the amount of evidence available. The less known about a given event, the more likely scholars (and others) will turn to climate change or environmental decay as explanation.

Even if the environmental interpretations of these collapses are right, their relevance to contemporary issues is modest.* The soci-

---

* These words were written before the publication of *Diamond* (2005) which argues the opposite for the Norse and the Maya.

eties in question were low-energy societies, using only their own muscles and, in the cases of the Indus and the Norse, those of their domesticated animals. They had very limited scientific knowledge— although certainly the peoples of the Indus and the Mayans knew a good deal about their immediate surroundings. Their technologies in every case were simple and evolved only slowly. For these reasons they seem, and they are, far-removed from today's industrialized, high-energy, high-technology world.

But modern history is another matter. Environmental considerations are directly relevant to today's issues, all of which grow out of them. The data are much better, although historians naturally differ on matters of interpretation. Caution of course is required: the future of global change, of politics, of technological evolution, is too much for human minds to grasp in the sense that we cannot confidently predict any of the things that matter most. We can understand their unfolding only when it is over with and in front of our eyes—and sometimes not even then. Even a precise sense of history will not allow us to build predictive models that integrate social and natural systems. This predicament calls to mind Churchill's aphorism about democracy, which he said was the worst of all political systems, except for all the others. History, one might say, is the worst of all possible guides to the future—except for all the others.

The first thing to understand is the bizarre, anomalous, and thoroughly unsustainable character of the recent past, especially the last half century. To this end, a glance at deeper history is especially helpful, as it shows quite clearly how peculiar modern times have been. Let us look at recent trends in population, economic growth, and energy in historical perspectives. The length of those perspectives will vary with the quality of the data.

II. Economic Growth Since 1 A.D.

Economic historians go to great lengths to measure the size of economies and their growth rates in times past, an exercise that is fraught with difficulties. The most sophisticated attempts, such as those of Angus Maddison,[3,4] whose figures appear in Table 1, recognize the difficulties and try to adjust accordingly.

Two thousand years ago, the size of the world economy was about $100 billion, roughly equivalent to the national economies of Nigeria, Finland, or Israel today. Over the next millennium it did not grow much (less than 0.01% annually) because population did not grow

Year	1 A.D.	1000	1500	1820	1870	1913	1950	2003
World GDP	103	117	239	694	1102	2705	5336	27076

Source: Maddison 1995, 2001

Table 1. World GDP, A.D. 1 to 2003 (in billions of 1990 dollars)

much and productive technology evolved very slowly. By 1500 the world's economy had attained a size roughly the same as today's Belgian or Philippine economy, thanks to a modest acceleration of population growth and a more modest one in technology and efficiency. After 1500 the world's markets grew more integrated, leading technologies were applied to more regions, and growth quickened so that by 1820 the world economy had reached a size equivalent to the economy of Canada, Spain, or South Korea today. Then a sea change took effect: with the Industrial Revolution humankind tapped energy stocks accumulated over eons in the form of fossil fuels. This allowed much faster and more reliable transport, much more thorough economic integration of disparate regions around the world, and all the rewards of specialization and exchange on an increasingly global scale. It also allowed a huge expansion in industries that hitherto had been held in check by energy or labor shortage. The exuberant expansion of the world economy reflects these trends. In the 180 years after 1820, the growth rate exceeded 2.2% annually, with the fastest periods coming, in declining order, in 1950–73, 1870–1913, and 1973–2001.

The world's economy in 2000 was about 270 times as large as in 1 A.D., and 125 times as large as in 1500. Almost all the growth took place after 1820, most of it after 1870. At all times the majority of the growth came simply from population growth. Income per capi-

Year	1 A.D.	1000	1500	1820	1870	1913	1950	1998
GDP per capita	444	435	565	667	867	1510	2138	5709
Closest national equivalents today	Sierra Leone	Sierra Leone	Tanzania	Mozambique	Chad, Rwanda	Togo, Sudan, India	Albania, Zimbabwe, Honduras	Turkey, Tunisia, Costa Rica

Table 2. World GDP Per Capita, A.D. 1 to 1998 (in billions of 1990 dollars)

ta grew much more slowly than the number of people did. Table 2 reveals the less spectacular trajectory of per capita world GDP over the ages.

Table 2 shows that while the world economy grew more than a hundred-fold between 1500 and 1998, on a per capita basis the growth came to less than ten-fold. We are, on average, about ten times richer than people who lived five centuries ago, and nearly four times richer than people were at the time of World War I. These figures are global averages and they conceal gigantic differences among regions, countries, and individuals. Nonetheless, this record of economic growth per capita, especially since 1820, must count as one of the signal achievements of the human race. It derives mainly from technical change in methods of production, allowing the use of massive stocks of energy in the form of fossil fuels, but also from the furtherance of market integration and the consequent efficiencies of specialization and exchange.

This extraordinary record of modern growth since 1820, and the dizzying pace of growth since 1950, came at a price; or, perhaps better put, at two prices. The first is that the growth was (and remains) sharply unequal, so that the modern age was (and remains) one of differentiation, of growing divides between rich and poor. This of course may be seen as a good thing because it means that some people, indeed more than ever before, live free from poverty. But if one prefers equality, it is decidedly a bad thing. At any rate, it contains the seeds of further instability because in the present age of cheap information, the world's poor will not remain ignorant of their plight, and many of them will not remain resigned to it. They will respond with ambition, hard work, migration, revolution, crime, and other forms of initiative, all intended to reconcile the yawning gap between their circumstances and their wishes. Growing economic inequality in an age of cheap information is a combustible recipe.

The second cost is environmental. Most of the human action that gets recorded as economic activity, and thus is reflected in GDP figures, has environmental consequences. Roughly speaking, the larger the economy, the greater the environmental disturbance. This, it must be stressed, is only roughly true. A great deal depends on the character of a given economy: farming and mining are much more environmentally disruptive than, say, accounting or teaching. Even what at first glance appears to be the same activity (for example, steelmaking) can have significantly different environmental consequences when done with 1890s technology as opposed to today's technology. Nonetheless, the huge modern growth in per capita incomes and associated consumption remains an anomaly of the last century, primarily of the last half-century, and explains, on a superficial level at least, the great environmental turbulence of our times. Let us move beyond this admittedly superficial level of estimated GDP figures.

III. World Population History Since 1 A.D.

People are much easier to count, and to estimate, than putative dollars of GDP. The following estimates are more reliable than those concerning the world economy, but those prior to 1900 are inevitably just estimates.[5,6,7,8] When humankind first invented agriculture, some 10,000 years ago, there were somewhere between 2 and 20 million members of our species roaming the earth. Farming allowed much denser populations than did hunting and gathering, especially with the advent of irrigation in favorable river valleys. By 1 A.D. we numbered perhaps 200 or 300 million, about as many people as the United States contains today. A millennium later, in 1000, world population was probably still below 300 million. By 1500, our numbers had climbed to around 400–500 million, and were growing by well under 0.1% per year, which was fast by previous standards but negligible by current ones. Famines and epidemics pruned back growth, and child mortality was such that half of all babies did not live to their fifth birthday. Details varied over time and from place to place, but this general regime of slow population growth interrupted by recurrent catastrophe prevailed in all agrarian societies.

In the eighteenth century this durable regime began to change. Death rates declined, at first slowly, and only in some parts of the world. Famines and epidemics became rarer, and child mortality diminished slightly too. This reflected the gradual microbial unification of the world, to the point where several deadly infectious diseases became endemic—not universally, but widely. It also reflected slightly better nutrition in many world regions. By 1820 human population reached a billion, and added another billion by 1920 or so. Now effective public health measures added to the prolongation of life, checking the spread of water- and insect-borne diseases. In a mere 40 years—by 1960—we passed the 3 billion mark. The improvements in health and life expectancy that initiated the modern rise in population now spread round the world, even to the poorest places where families had long compensated for high infant and child mortality by having many babies. Vaccinations and antibiotics helped bring about this health revolution. So after 1950 a true population explosion began, bringing human numbers to 4 billion by 1975, 5 billion in 1987, and 6 billion in the year 2000. In this remarkable span of time, from 1950 to 2000, world population grew at rates roughly 10,000 times as fast as for most of human history (that is, the pre-agricultural era), and at rates 50–100 times as fast as during the preindustrial millennia. If the contemporary rates had been in force since the dawn of agriculture, the planet would now be the core of a squiggling mass of human flesh thousands of light-years in diameter and expanding with a radial velocity faster than the speed of light![9]

These rates, one may safely assume, will not last long. Indeed, demographers now expect that human numbers will peak in half a century or so, at about 9 or 10 billion. The demographic pattern of the recent past is not only bizarre in light of the past, but will one day appear so in light of the future as well. Whether global population will stabilize at 9–10 billion, or enter into an era of decline or of unstable oscillations, is anyone's guess. We may safely say, however, that as with economic history, so with demographic history: the period since 1950 has been, in the perspective provided by longer sweeps of time, unimaginably bizarre.

IV. Energy History

The energy history of the last two centuries, and especially the last half century, was also a remarkable departure from more durable patterns of the past. For most of human history the only energy available for getting work done was muscular energy, derived from the chemical energy stored in edible plants, which in turn came from energy flows from the sun captured by photosynthesis. Our distant ancestors often ate meat, which allowed them to supplement the fuel supply to their muscles with chemical energy stored in animal tissues, itself ultimately derived from plants and photosynthesis.

With the domestication of draft and pack animals, a slow process that began not long after the first plant domestications, the total energy available for work expanded somewhat. Horses, oxen, water buffaloes, or llamas could be trained for useful work and sometimes could draw their energy from plants that human beings could not eat. Further, with the inventions of sails, windmills, and watermills, another small windfall of energy became available. Agrarian societies making use of all these inventions, say Japan or France in the year 1700, harnessed four to six times as much energy per capita as did hunting and gathering populations.[10]

But their energy harvest remained almost entirely confined to what they could capture from the annual flows from the sun, and their methods of capture were extremely inefficient: in each conversion process, from sunshine to plant tissue and from plant to animal tissue, the great majority of energy was lost. Agrarian societies could not store energy conveniently, except potential heat energy in the form of firewood or charcoal. This energy inefficiency put a sharp limit on what agrarian societies could do. The vast majority of human beings in this situation could only be poor, winning their precarious subsistence through grinding toil. Dozens or hundreds of people had to work hard and suffer some of the fruits of their labor to go to someone else to make anyone leisured or (by modern American standards) anything close to comfortable.

Fossil fuels and steam engines changed all that, beginning around 1800. Preindustrial societies used wood and charcoal for heat energy to keep bodies warm in winter, and to fuel enterprises such as glassmaking or beer brewing. In wood and charcoal they were tapping energy supplies that represented scores or (rarely) hundreds of years of accumulated solar energy. But coal stored eons' worth of solar energy in concentrated form. With coal and steam engines, humankind shattered the old energy constraints on society and suddenly could do far more work and create much more wealth. Modern industrial societies capture and use far more energy per capita than did agrarian societies. This meant that for the first time poverty and grinding toil were no longer mandatory. The world used about five times as much energy in 1900 as in 1800.

Then the real boom began. Oil contains roughly twice as much energy per ton as does coal, and can be used in a wider variety of applications, especially in transport. World oil production was negligible before 1880, considerable by 1920, and huge by 1970. Oil accounts for the greater part of the tremendous expansion in energy use in the 20th century, which was perhaps thirteen- or fourteen-fold. Most of that growth took place after 1950, when the biggest Middle East oilfields came on stream and ordinary people in Europe and Japan followed the North American precedent and took to automobiles. Like the simultaneous growth of economies and populations, the boom in energy consumption was unevenly distributed around the world and the global averages hid sharp inequalities. In the richest societies, the average person had (by 2000) the equivalent of about 75 "energy slaves" (meaning energy equivalents of the work a human being is capable of performing) working around the clock at his or her disposal. In the poorest societies, the average person had one energy slave. Indeed, it was precisely because the average Ethiopian or Bangladeshi had to do all his or her work, without much use of inanimate energy, that they were so poor.[11, 12]

Variable	Co-efficient of Increase 1890s–1990s
Human population	4
Urban proportion of human population	3
World Economy	14
Total urban population	14
Industrial output	40
Energy use	13–14
Coal production	7
Water use	9
Irrigated area	5
Cropland	2
Forest area	0.8 (20% decrease)
Bird and mammal species	0.99 (1% decrease)
Fin whale population	0.03 (97% decrease)
Marine fish catch	35
Cattle population	4
Pig population	9
Carbon dioxide emissions	17
Air pollution	2–10
Sulfur dioxide emissions	13
Lead emissions	8
	Source: adapted from McNeill 2000.

Table 3. Magnitudes of Global Environmental Change, 1890s to 1990s

V. Environmental Trajectories

The period since 1900, and especially since 1950, has been one of lightning change and exuberant expansion in terms of economic growth, population, and energy use. Mainly for these reasons, it has been just as turbulent in its environmental history. Table 3 attempts to give some precision to the magnitude of some environmental changes in the twentieth century, and to some of the driving forces behind them. It is, of course, selective. It ignores variables for which the co-efficient of increase would be astronomical (the number of automobiles or of organic chemicals) and those for which it would be infinite (chlorofluorocarbon emissions or tractors).

The table shows the dramatically unstable character of the modern relationship between humankind and the rest of nature. Driven by energy use, population growth, urbanization, and technological change, among other things, we have lately created a regime of perpetual disturbance in global ecology. In such unstable circumstances we have prospered mightily as a species, growing in number, in wealth, and in the share of the biosphere's energy and materials that we can turn to our own uses. It is as if we had craftily shaken things up so as to help ourselves at the expense of other species. But the regime of perpetual disturbance is not a grand global plot; rather, it is the unintended consequence of the strivings of billions of people seeking traditional goals of security and well-being. In recent times the combined impact of their efforts was magnified by their growing numbers and their growing technological expertise.

VI. Ideas and Politics:
The More Things Change the More They Stay the Same

In his famous novel *The Leopard,* Giuseppe di Lampedusa has one aristocrat say to another, à propos of their ambition to stay atop the social order in Sicily in a time of social revolution: "If we want things to stay as they are, things will have to change." If one wanted global ecology to stay the same, then prevailing ideas about the ideal role of the state and the ideal trajectory of the economy would have to change. The regime of perpetual disturbance has drawn some of its impetus from the overlapping worlds of ideas and politics. The dramatic changes in the human position within the earth's ecosystems has scarcely registered in either world, which is why they continue to add to the ecological tumult: global ecology changes as rapidly as it does in part because ideas and politics, from an ecological point of view, change so slowly.

Ideas and politics in the 19th and 20th centuries shifted and shifted again, in response to the vast social and economic changes afoot, and in some measure helped drive and direct those changes. But in a very fundamental sense they did not change much. Conceptions of the role of nature in human affairs, the role of humans in the natural world, and of policies appropriate for maintaining these roles—all this showed remarkable stability in the midst of great flux.

First, we must recognize that the ecologically important ideas and policies are not the explicitly environmental ones. Since the late 1960s, environmental movements have sprung up—more noticeably in some countries than others, but almost every country has seen fit to establish a ministry or department for environmental affairs. Yet the views of environmentalists have yet to compete seriously in the marketplace of ideas with reigning notions about appropriate human behavior. Instead, it is the big ideas—ones accepted by many millions of people about how society ought to be organized—that carry big environmental consequences. And real environmental policy is nowhere made by the ministries and departments of the environment. Instead, it is made by the traditionally powerful organs of government: the ministries of finance, industry, defense, and agriculture. They have made, and continue to make, environmental policy as an accidental by-product of their efforts to pursue their normal priorities.

Consider three big ideas of the twentieth century: nationalism, communism, and the "growth fetish." They are all related, all big ideas about the appropriate structure, principles, and priorities of society. Modern nationalism grew out of the ideologies advanced by the French Revolution and prospered in the 19th and 20th centuries under the impacts of migrations, public education, colonialism, and other forces.[13,14] At times, nationalist sentiment helped promote nature conservation as citizens concluded—usually in times of rapid urbanization and industrialization—that in preservation of the landscape, one could safeguard the essence of German or English or Japanese identity. The Nazis, for example, were keen on nature preservation (in principle more than in practice) in the belief that the true German character was connected with primeval woodlands and peasant farming.

In other settings, nationalism served as a force for radical ecological change. In Amazonia after 1960, Brazilian governments sought to 'Brazilianize' the vast tracts of rain forest within their borders (indigenous peoples did not count as Brazilians). To achieve this goal they built highways and feeder roads, encouraged settlement by "real" Brazilians, and subsidized ranching and other activities that involved cutting and burning off some 10–12% of the world's largest tropical forest within the next 40 years.

Indirectly, nationalism had further environmental effects through policies of pro-natalism, with which dozens of rulers sought to realize nationalist goals by raising population levels, which, in turn, raised the level of human ecological impact. The most successful pro-natalism took place in Romania, where in the mid-1960s the dictator Nicolae Ceaucescu set a growth target of 30 million Romanians by the year 2000. He outlawed all manner of birth control and set his formidable secret police the task of ensuring that Romanian women did not shirk their reproductive duty. In 1966–67 the birth rate doubled within a year.[15,16] The most consequential pro-natalism was that of Mao Zedong in China. Mao generally (not always) took the view that there could never be too many Chinese. He discouraged the population control efforts that several Chinese leaders, including his closest lieutenants, favored. In collectivizing agriculture (from 1953) he inadvertently removed village and family-level control over fertility, provoking a baby boom that lasted into the 1970s. Mao of course did not intend collectivization as a pro-natalist policy, but he welcomed the effect and resisted policies to counter it. China doubled its population in the Mao years (1949–76), bringing on acute problems with respect to fresh water supplies, urban air pollution, desertification, and forest decline. Only after Mao's death, and after the addition of 500 million more Chinese, did the state embark on a serious population control effort. By that time the situation seemed so dire that China enacted the most draconian family limitation policy in world history, the one-child policy, and enforced it fairly successfully.[17,18]

Communism was another big idea of the twentieth century that spread successfully around the world. Part of its success in Russia, China, Vietnam, and Cuba lay in the fact that it promised to deliver a fundamental nationalist goal, independence from foreign domination. In any case, deep within Marxism was the ambition to dominate nature, on the theory that when man could exploit nature sufficiently he would no longer exploit his fellow man. In the Soviet Union this became a major tenet of the state dogma, as expressed by the writer V. Zazurbin in 1926:

"Let the fragile green breast of Siberia be dressed in the cement armor of cities, armed with the stone muzzle of factory chimneys, and girded with iron belts of railroads. Let the taiga be burned and felled, let the steppes be trampled. Let this be, and so it will be inevitably. Only in cement and iron can the fraternal union of all peoples, the iron brotherhood of all mankind, be forged."[19]

The enthusiasm for the transformation of the environment, for "correcting nature's mistakes" as Soviet planners often put it, became central to Soviet policy from the 1930s. The Soviets built huge dams and long canals. Their attachment to giantism, in agriculture and industry as well as in architecture and statuary, led them to reorganize farmers' fields into huge collective farms, with fields that stretched for miles, inviting wind and water erosion. Since nature had not blessed the U.S.S.R. with good land for cotton production, after 1960 planners diverted the waters that fed the Aral Sea to grow cotton in Central Asia, eventually strangling the Aral, killing off its teeming fisheries and canning industry. As the Sea shrank to half its size, steppe winds lifted salts from its exposed seabed and deposited them throughout Central Asia, damaging cotton and other fields alike. The assassination of the Aral Sea was perhaps the greatest environmental disaster of the 20th century. It was also deliberate: the engineers involved knew the sea would dry up, but reckoned it was worth doing for the cotton, which proved so poor in quality as to be unsaleable outside the Soviet bloc.

Soviet communism was environmentally important also because it resisted technological change. As late as the 1980s the Soviets used the open-hearth steelmaking process, invented in the 19th century and long obsolete in Japan, South Korea, Western Europe, and North America. Factory bosses had quotas to meet and no incentive to experiment with new technologies, leaving the U.S.S.R. with an energy-guzzling, pollution-intensive economy until the very end—which fact helped bring about the end of the U.S.S.R. when dissidents and other citizens objected to the many forms of environmental degradation in their homeland.[20,21]

Ideological commitments in China helped make that country one of the most polluted in the late 20th century as well. Industrial production, billowing smokestacks, and the like acquired totemic meaning in Mao's China, and the resulting air and water pollution seemed, to the leadership at least, a small price to pay for the iron

and steel. During the Great Leap Forward (1958–60), industrialization followed a particularly resource-inefficient path, as Mao
exhorted peasants to produce steel in backyard furnaces, which
required converting all available wood into charcoal. As in the
U.S.S.R., the unresponsiveness of the leadership to citizen com-
plaints meant that pollution problems rarely acquired sufficient priority in the councils of state to produce effective anti-pollution
action. Indeed, for decades the official position in several communist countries was that pollution could exist only in capitalism.

Communism shared one of its strongest ideological commitments with capitalism: the growth fetish. This was a flexible and
seductive creed, appealing to almost everyone in authority, because
economic growth hid a multitude of political sins. Populations
would put up with corruption, vast inequalities, and repressive
police states if they were confident that in the years to come they
would be materially better off than at the moment. After the dislo-
cations of the depression of the 1930s, consistent economic growth
became the holy grail of political economy. During WWII economists displayed such skill in managing the American and British war
economies that they became high priests in the postwar world,
entrusted with the sacred duty of manipulating aggregate demand
so as to deliver full employment and steady growth. The international power and prestige of the Americans in the postwar era
helped spread this gospel far and wide. Within its sphere, the
U.S.S.R. promoted a vision of growth administered more by engineers than economists, that would, in theory, beat the capitalists at
their own game.

The American economists had much the same approach to the
environment as the Soviet engineers. Discarding earlier views that
took nature seriously (if simplistically), such as those of Malthus or
Jevons, after the 1880s formal economics evolved into a bloodless
abstraction in which nature figured only as a storehouse of
resources waiting to be tapped. Infinite substitutability meant that,
as one American Nobel laureate put it, the economy could in effect
get along without natural resources. Growth could last until the
extinction of the sun, some 5 to 7 billion years hence.

The growth fetish derived from an age with empty land, plentiful
forests, vast shoals of undisturbed fish, a robust ozone shield, and,
in general, intact ecological buffers. It helped pare those buffers
down. The idea then outlasted its usefulness, surviving into an era
of greater ecological stress. This is standard: successful ideas

become dogmas; they mesh with social and political institutions; they become, in a word, locked in. Successful technologies do the same, limiting future options. Economic thought, in the capitalist world as much as the communist one, did not adjust to the new conditions it helped to create. Instead it continued to legitimate—and in an indirect sense, to help cause—massive ecological changes.

Nationalism, communism, and the growth fetish were all successful ideas in that they acquired millions of supporters and became central to the policies of many states. Thus their ecological implications extended widely across the earth. While communism is at least temporarily dead as a big idea, the traditional emphases of nationalism and the growth fetish remain with us in the 21st century, and will continue to help fashion new dramas of ecological change.

VII. Conclusion

It is natural for people to assume that the world as they find it is normal, to draw conclusions on the basis of their own experience and observations. But at present this natural assumption is profoundly misleading. Those whose experiences and observations are confined to the past 100 years—or worse yet, the past 50 years—know only the most unusual era in the environmental history of the world. If we lived 700 or 7,000 years instead of a mere three score and ten, we would understand this readily on the basis of memory alone. As it is, we need a longer historical perspective to recognize the peculiarity of our times, and to appreciate the radical unsustainability of our current ways.

We are, beyond a shadow of a doubt, doing what Einstein said God would not do: playing dice with the world. Recasting the planet's biogeochemical flows, especially of carbon and nitrogen, is no small feat. Nor is joining the editorial board of biological evolution, helping select (normally unconsciously) species for extinction or survival. It is possible that we can do such things indefinitely with no great ill effects for the human species. It is possible (although I think unlikely) that we can pursue the present unsustainable ways for a century or two, or even three, at little cost. It is possible that our current unsustainable regime can be replaced by another one, giving us a new lease on ecological life. Imperial China, for example, followed one unsustainable path after another for 3,000 years,[22] and its successor state, the People's Republic of China, has pursued an even more radically unsustainable path for more than

50 years. But the fact is we cannot know how long we can act this way, or how unwelcome the effects might be if we continue too long.

The historical perspective can make clearer, perhaps even make obvious, the ecological peculiarity of our times, so that one need not be among Machiavelli's men of talents to see it. It can, perhaps, help generate sufficient foresight so that evils which might arise are discovered in time, perceptible to everyone before they are suffered to increase to such a height that there is no longer any remedy. Finding and administering a remedy—treating evils upon a proper principle, as Machiavelli put it—will perhaps be intellectually, although not politically, a trifle easier.

Notes

1 McIntosh, J. 2002. *A Peaceful Realm: The Rise and Fall of the Indus Civilization.* Boulder, CO: Westview.
2 Webster, D. 2002. *The Fall of the Ancient Maya.* London: Thames and Hudson.
3 Maddison, Angus. 1995. *Monitoring the World Economy, 1820–1990.* Paris: OECD.
4 –2001. *The World Economy: A Millennial Perspective.* Paris: OECD.
5 Cohen, J. 1995. *How Many People Can The Earth Support?* New York: Norton.
6 Livi-Bacci, M. 1992. *A Concise History of World Population* Oxford: Blackwell.
7 Maddison, Angus. 2001. *The World Economy: A Millennial Perspective.* Paris: OECD.
8 Caldwell, J.C. & T. Schindlmayr. 2002. Historical population estimates: unraveling the consensus. *Population and Development Review* 28: 183–204.
9 Cipolla, C. 1978. *An Economic History of World Population.* Harmondsworth: Penguin Books.
10 Sieferle, R. 2001. *Der Europäische Sonderweg: Ursachen und Faktoren.* Stuttgart: Breuninger Stiftung.
11 Smil, V. 1994. *Energy in World History.* Boulder, CO: Westview.
12 McNeill, J.R. 2000. *Something New Under the Sun: An Environmental History of the Twentieth-Century World.* New York: Norton.
13 Wiebe, R. 2002. *Who We Are: A History of Popular Nationalism.* Princeton: Princeton University Press.
14 Roshwald, A. 2001. *Ethnic Nationalism and the Fall of Empires.* London: Routledge.
15 Kligman, G. 1998. *The Politics of Duplicity: Controlling Reproduction in Ceausescu's Romania.* Berkeley: University of California Press.

16 Chesnais, J. 1995. *Le Crépuscule de l'Occident: Démographie et Politique.* Paris: Editions Robert Laffont.

17 Lee, J. & W. Feng. 1999. *One Quarter of Humanity: Malthusian Mythology and Chinese Realities.* Cambridge, MA: Harvard University Press.

18 Shapiro, J. 2001. *Mao's War Against Nature: Politics and The Environment in Revolutionary China.* New York: Cambridge University Press.

19 Hillel, D. 1991. *Out Of the Earth: Civilization and the Life of the Soil.* Berkeley: University of California Press.

20 Feshbach, M. & A. Friendly. 1992. *Ecocide in the U.S.S.R.* New York: Basic Books.

21 McNeill, J.R. 2000. *Something New Under the Sun: An Environmental History of the Twentieth-Century World.* New York: Norton.

22 Elvin, M. 1993. Three thousand years of unsustainable development: China's environment from archaic times to the present, *East Asian History* 6:7–46.

14

Governance Barriers to Sustainability

Richard D. Lamm

> *You would be surprised at the number of years it took me to see clearly what some of the problems were which had to be solved .... Looking back, I think it was more difficult to see what the problems were than to solve them.*
>
> —Charles Darwin

Survival research in the area of governance raises the question of whether our governmental institutions are equal to the magnitude of the problems we face. This is always a challenge to government; "the hole and the patch must be co-terminus," exclaimed Abraham Lincoln. Survival research raises basic issues upon which the dialogue is not only embryonic, but generally non-existent. Can we not only correctly identify the challenges facing humankind, but also get their remedies through the democratic process? Can democracy adequately anticipate the new set of survival problems we face?

The Culture of Growth Vs. the Culture of Limits

Sustainability will require a basic change in the culture of the developed (and underdeveloped) world. C.P. Snow, in his famous talk, *Two Cultures,* contrasted the differences between the world of science and the world of letters and went on to observe, "Between the two, a gulf of mutual incomprehension … sometimes hostility and dislike, but most of all, lack of understanding."[1]

This same "two cultures" metaphor is useful to spotlight what I consider a new chasm of "mutual incomprehension" that the world must soon resolve—the irreconcilable conflict between the culture of growth and the culture of limits. Are resources finite or infinite? Can we solve the problems of growth with more growth? Will existing mechanisms and institutions (including capitalism) be sufficient and successful for the next 200 years as they have been for the last 200 years? There is the culture of growth which denies limits, and the culture of limits which seeks to accept and adapt to those limits. These two cultures hold opposing worldviews. They urge conflicting courses of action and accuse each other of myopia or worse. It is often a dialogue between the blind and the deaf.

Most of human experience is on the side of both continued population and economic growth culture. The world of growth has succeeded brilliantly. It allowed survival in a harsh world. It has brought health, wealth, increased life expectancy, leisure, and most important—freedom. Growth has approached the status of a religion. Sociologist Peter Berger points out:

> "Development is not just a goal of rational action in the economic, political and social spheres. It is also, and very deeply, the focus of redemptive hopes and expectations. In an important sense, development is a religious category. Even for those living on the most precarious margins of existence, development is not just a matter of improved material conditions; it is at least also a vision of redemptive transformation."[2]

But even in our religious fervor, we must ask, "can it last?" Is this a sustainable vision? Is growth the permanent secret to success for societies?

I would argue that for all our genius, we can't escape ecological limits. I think the greatest challenge of the international community is to modify, and in some cases reverse, mores and cultural attitudes that have worked well and under which we have prospered for hundreds of years. We can delay, but not totally avoid, the consequences of infinite human demands on a finite earth. A very fundamental New World has emerged—a set of circumstances which are as important as the industrial or agricultural revolution. It is to change the world of growth into the world of sustainability.

Growth is likely embedded in our DNA. It is not something we should easily give up. It has led humankind from the cave to our current standard of living. However, that is not the end of the argument, for as the stock market warns us: "Past success does not guarantee future success." The culture of growth has clearly served us well, but it begs the question before us: is it sustainable?

One of the human dilemmas is that we often see the world not as it is, but as we think it is. Columnist and thinker Walter Lipmann once warned: "At the core of every moral code, there is a picture of human nature, a map of the universe, and a version of history ..."[3] Our economy, our ethical standards, and our moral standards depend on the mental map we have of the world. Author Thomas Sowell points out that people have very different visions of how the world works. "Visions are foundations on which theories are built," and Sowell observes that most of us have mental maps of the world in our minds which do much to control our viewpoints and thus our policies.[4]

Are there limits in our physical world, or are those "limits" only limitations of our vision, creativity, technology, and ingenuity? Are there limits to human development in the physical world around us, or only in our minds? Can the mental map that Western Civilization has formed in our minds and human expectations be achieved in the physical world we live in? Is the past a guide to the future, or is it a "cultural trap" that keeps us from recognizing that the world is approaching carrying capacity? Could we end up being victims of our past successes because we have inherited the wrong mental map?

The jury's still out—neither side can claim victory, but the world is presently developing and increasing population and standards of living, and thus the burden of proof should be on those who seek to change what has worked so well. The culture of growth has served us well. The growth ethic is not something we should easily give up.

One of the great challenges of history is to know when a new world or new paradigm has emerged. It is my passionate belief that economic practices cannot be at variance with ecological reality. Our economic system must adapt to our ecological system or, at a minimum, our economic system cannot destroy our ecological system.

The assumptions that undergird our whole industrial society incorporate infinite resources. The culture of growth confidently feels that there are no limits that cannot be overcome. Clearly, public policy and most of our institutions, as presently structured, assume unlimited resources, infinite wealth creation capacity, and no global ecological limits.

I think the future can be better planned for by confronting limits to the best of our ability and heeding the warning that infinite growth cannot take place in a finite world. The fact that we have been so successful in pushing back those limits does not dissuade me from believing that those limits are real. "All modern day curves lead to disaster," warns former French President Valerie Giscard d'Estaing. (Population, consumption, environmental destruction.) Human civilizations are presently living on the upper shoulders of some incredibly steep and unprecedented geometric curves. They cannot continue indefinitely. No trees grow to the sky.

I believe that we are surrounded with evidence that increasingly shows that something is fundamentally wrong with the growth paradigm. Our globe is warming, our forests are shrinking, our icecaps are melting, our coral is dying, our fisheries are depleting, our deserts are encroaching, our finite water under more and more demand. I suspect these to be the early warning signs of a world approaching its carrying capacity. We cannot merely call upon human ingenuity, science and technology to develop new solutions to these new challenges. Technical know-how is not wisdom: it is important, but it is not enough. I once knew a man who knew seven languages and he had nothing to say in any of them. We must be more than technically proficient; we must instead change our mental map of the world, our culture, and our economy.

Genetic Barriers to Sustainability

Is the growth ethic genetic? Are governmental institutions helpless in face of genetically programmed traits that are leading us to destruction? There is a fascinating developing issue concerning whether human's genetic heritage—our "neuronal heirlooms"— are so deeply imbedded that humans cannot and will not stop themselves from overrunning the environment. The same genetic values that allowed *Homo sapiens* to survive and prosper in our harsh and unforgiving climate may be counterproductive to the earth we have subdued.[5]

"Is it possible that this successful species of ours also embodies one or two heritable flaws against which we have no defense, flaws embedded in our wiring of our brains or hidden in the coils of our massive DNA?"[6]

Paul Shepard, in a series of books and papers culminating in *Coming Home to the Pleistocene,* raised similar issues speculating that our essential human nature is a product of our genetic heritage, formed through thousands of years of evolution during the

Pleistocene epoch, and that it is this genetic heritage that lies at the heart of today's ecological and social ills. Can the rational offset the genetic? Careers will be won and lost over this and related issues.[7]

Institutional Barriers to Sustainability

An alternative field of inquiry is into the institutional barriers to sustainability. This inquiry would postulate that the problem is not in our genes, but in our institutions and hubris. The ancient Greeks believed the greatest tragedies come when well-meaning, intelligent, caring humans bring on disaster because they can't see the big picture or understand new circumstances. A chorus warns them of impending disaster, but they do not hear. Are we listening to the increasing chorus of opinion that warns us that our human institutions are inadequate to the task of sustainability? And if some are listening, can they effect change through the human institutions we have developed?

There are some institutional governance barriers that have long been debated. They not only relate to economic growth and population growth, but to all mega-problems which face a democratic political system. I would like to paraphrase Professor Kenneth Bounding, who said the modern public policy dilemma is that all our experience deals with the past, yet all our problems and challenges deal with the future. Can we really anticipate the magnitude of the problems and solutions that it will take to get to a sustainable future? Public policy too often blindly assumes that the future is an extrapolation of the past. New worlds always catch us by surprise. Barbara Tuchman, in her book, *Practicing History,* warns: "Policy is formed by preconceptions and by long implanted biases. When information is relayed to policy makers, they respond in terms of what is already inside their heads and consequently make policy less to fit the facts than to fit the baggage that has accumulated since childhood."[8]

Clearly this point applies to all governments. It is useful, however, to look at the United States' institutions and cultural biases as an example of the problems all democratic governments face. Human institutions are built on the foundation of our assumptions. We all have a mental map of how the world works, what works in that world and what doesn't work, what is functional and what is dysfunctional. Humankind has made mistakes in our assumptions as Copernicus, Darwin, and Freud have shown us. But those are easily identified exceptions. The examples are true, but too facile. Generally our public policy assumptions have been reliable, or at least workable. While there have always been those who say the world is going to end tomorrow, it's still here, we're still here, and the standard of living is going up in most of the world. We should never challenge existing assumptions lightly. Will and Muriel Durant, after studying a lifetime of history, observed: "99 out of every 100 new ideas that come down the road are bad ideas."

That is probably close to correct. We are, of course, right to assume. We must have some ground to stand upon. We can't live our lives or develop public policy without the foundation of some assumptions. But we are wrong to blindly assume what once worked will always work. History is always playing tricks on humankind.

I suggest four "preconceptions" that relate to the inability of the United States to come to grips with challenges which face it:

1. that the United States has a divine destiny;
2. that our problem solving machinery and institutions are equal to the magnitude of the problems we face;
3. that our political system and democracy are sustainable;
4. that our belief in growth of population and the economy is an unquestioned civic good.

I raise these issues not to refute the assumptions, but to suggest that we should more fully debate them and others like them. A prudent society will from time to time examine its assumptions to judge their continuing validity.

Assumption I: that America has a divine destiny or, put another way, that God is an American. Some scholars call this American exceptionalism, the belief that the U.S. is especially blessed and, as the old saying goes "God watches out for drunks, fools and the United States of America." This strong belief in our divine destiny runs very deep in our psyches.

People believe too easily that their nation has divine blessing and direction. Similar assumptions have often been made, but have never yet been true—in all of history. All great civilizations have believed some variation on this theme. These assumptions stanch our sense of urgency, or they take problems as part of God's plan we are helpless to correct. If environmental degradation is simply part of God's plan or his punishment, it is hard to rally support to correct it. Greatness is not a divine guarantee, but a continuing challenge. The United States is a unique and wonderful place, but so were Greece, Rome, and the British Empire. "The sun," it was claimed, "would never set on the British Empire," but it did set. No country's mental map should contain a belief that they have a special dispensation from history's forces. We shall always have to earn our prosperity and our greatness.

Assumption II: that our governmental institutions are equal to the magnitude of their problems. It is imperative that nations have institutions efficient and effective enough to solve the problems they face. Related to this is the feeling that we can solve our national problems within the existing political dialogue. I am very concerned lately whether our problem solving machinery is adequate. Do we have the institutions which will allow us to make changes of the magnitude required?

Our very past successes might set us up for failure. Our nation and our system have been pretty good at self-correction. We always seem to correct our excesses. Can we do this again? Perhaps. Hopefully! But, we face a new set of particularly difficult problems as we move to sustainability. Can the governmental institutions which distributed a growing economic pie turn around and achieve and administer sustainability? I suggest that government generally has structural problems that cannot be solved with "politics as usual." As political scientists Levergood and Breyfogle point out, "We must realize that our current crisis of self-interested bickering and anarchy derive neither from our own selfishness, nor from the dishonesty and incompetence of politicians, but rather from political institutions that are no longer able to restrain the worst within us."[9]

An example of this is the way that campaigns are increasingly being run in democracies. Campaign reform is imperative and overdue. A giant "For Sale" sign hangs over America's political system. The biggest correlation of victory in politics is not your issues, your character, nor your energy; it is how much special interest money you have been able to collect. It is not the ideas in your head, but the money in your pocket. Voter cynicism and alienation are at all time highs, and cynicism and alienation goes right to the heart of a democracy.

Public alienation from the political system is the greatest danger any government (especially a democracy) can face. Columnist E.J. Dionne has observed, "A democracy which hates politics cannot remain as a democracy." We must act to restore the public confidence and rescue our nation from factionalism and cynicism. I increasingly do not believe this can be done within the existing two-party system.

Power tends to corrupt all human institutions, but the U.S. Constitution and its balance of powers have worked well to deal with corruption or abuse of power. The two-party system plays a real role in this process in debating issues and exposing self-interest. To a remarkable degree, we have enjoyed a self-correcting system. Now, however, the special interests have found a way to avoid this self-correcting system. They have taken over both political parties. They advance their coercive agenda by electing all or most of those who make the rules. It does not matter whether someone is a Republican or a Democrat—do they support and defend your particular interests? Special interests can often pass and stop legislation at will. Their actions are becoming increasingly blatant. Lobbyists' campaign checks are being handed out on the floor of Congress. What better symbol of how corrupt the process has become? It isn't the validity of your cause, but how you work the system. We use the resources of the uncomplaining many to satisfy the complaining and self-interested few. Eventually, this is a political Ponzi scheme that is bound to crash.

This dilemma seems to be beyond the reach of the normal political process. The U.S. Constitution is still an effective document. We still have two major political parties. The problem is that there has developed an end-run strategy which controls public policy through election laws, party rules, and special interest money. Our problems are more with our funders than our founders. It is not our Constitution that is flawed, but this end-run strategy that has developed to elect, fund, and lobby all our elected officials. We cannot express the national will because we cannot get around those with a special agenda. It is Thomas Jefferson's ultimate nightmare: Alexander Hamilton's "economic elite" has taken over both political parties.

The first step in solving a problem must be to correctly identify the nature of the problem. We have an institutional problem more than a political problem. Both political parties are for sale; both are hopelessly compromised by special interest money. Our political institutions, instead of being part of the solution, have become part of the problem. Only a new party, unencumbered by the past, can take the money out of politics or reduce its caustic influence. Real political reform may require a Constitutional amendment allowing limits on campaign spending, or possibly publicly funded campaigns, and nothing short of a political revolution will garner the needed political support. The Republican and Democratic foxes will never adequately protect the henhouse. The historic solutions have themselves become the problem.

There is an important distinction here. There will always be special interests; they are not inherently evil. Madison, in *Federalist No. 10,* observed they were integral to democracy. However, when they change their modus operandi from argument and logic to buying political influence, it is time to act. They must inevitably exist; they do not have to inevitably control the process.

There is a natural coalition out there for a new political movement. Take the pro-choice Republicans and the economically realistic Democrats and you have the core of a new party—the fiscally responsible Democrats and the socially progressive Republicans. Lop off those who have no faith in government and those with a blind faith in government, and organize the remainder. Most Americans neither want to dramatically extend the power of the federal government, nor dismantle it. The center of American political opinion is ripe for conversion.

Assumption III: that democracy is capable of solving all problems, including the transition to sustainability. Democracy has served us well, and none of us can think of a system that has been as successful for as long. It is not clear to me, however, that democracy is history's final governing system, or that it can distribute pain or sacrifice if that becomes necessary. Is it flexible enough to deal with long term problems? Can democracy act in society's long term interests? I worry about whether democracy can prosper in an age of self-gratification and civic ignorance.

The famous Danish physicist Niels Bohr once said: "The opposite of a truth is a lie. The opposite of a profound truth may well be another profound truth."[10] Of course democracy has proven itself the best system of self-government. But is it sustainable?

We have built a system without brakes that relentlessly pre-spends our children's money. It's the public policy equivalent to binge shopping—the gratification is so immediate and so tangible, the price deferred.

Successful democracy, in the long run, requires some allegiance to and respect for the future. In the short run, modern public financing gives public policy makers incredible opportunities to encumber the future for present political gain. Running deficits is the most obvious one, but there are many others. One can usually start a popular new program or give tax relief by under-funding pension systems, or inadequately maintaining infrastructure, or privatizing a government function, or selling off assets. The list is long.

The best politics these days seems to be the worst long-term public policy. The real test of our institutions is their ability to handle hard times. Does not democracy require moral, educated, and hard-working citizens? (John Adams once said "Our system of government was only made for a moral and religious people, and it is inadequate for any others.") I suspect that it is too early to claim permanent victory for democracy.

Democracy seems to be a reactive system. Where were Hitler's armies when Great Britain turned to Churchill? Answer: Dunkirk!

I am convinced that my generation created expectations that made it profitable to behave in the short term in ways that were destructive in the long term. Benefits became entitlements. Recipients became constituents. Privileges became rights.

How do you downsize expectations in a democracy? What unions call "givebacks"? We can't have Democratic spending policies and Republican taxing policies. Democracy requires citizen participation and some level of trust in the system. Today we have built a reciprocal distrust of the system. The public looks at politicians as crooks and self-serving liars. Simultaneously, politicians look at the public as sheep to be manipulated by 30-second spots. This is not the way Madison and Jefferson envisioned.

Deficit spending is worldwide and at every level of government. It's Faustian. Politicians can successfully run for office and promise more spending and tax relief—the only price we pay is the future of our children. A 200 year ethos that budget deficits are only appropriate during times of war or recession has been completely eroded to be replaced by a new and, so far, successful strategy of using deficits to avoid making unpopular decisions. "Need," not "affordability," will drive governmental policies as long as that government has a way to transfer the costs to the future.

My generation has been one of "credit card liberals." Keynesian policies require symmetry of application: deficits to stimulate the economy in bad times, and surpluses to avoid inflation during good times. There are two wheels on the bicycle, and both are indispensable. Either one alone will be ruinous.

We have been consuming wealth that rightly belongs to our children, without any hint of justification. We live in the afterglow of

The New Deal. What worked when we were the world's largest creditor nation, with a small federal debt to fight WWII and a depression, cannot continue to fund consumption for the world's largest debtor nation. American political culture makes it politically irresistible to be a good neighbor, but almost politically impossible to be a good ancestor.

Assumption IV: that economic growth will always fit within the ecosystem. Economic growth has also served us well. It has been the secret of our success—it has increased our standard of living worldwide. We have always assumed that the economic system and the ecological system are and have been separate issues. Will they continue to be? We must consider the possibility that the well being of each individual and each nation might be destructive to the ecosystem and thus help destroy the whole of which each is a part. It has not dawned on most of us yet that if it is a world of limits, there is a conflict between the welfare of the individual and the welfare of the whole.

E.B. White once said: "I own one share of corporate earth—and I'm very uneasy about the management." A number of new questions of public policy are being raised. What is a country's demographic destiny? Can the world sustain the standard of living that all countries are striving for? Can we have infinite growth in a finite environment? Is humankind losing its margin of error and setting itself up for some calamity? We never raised the issue before because we didn't recognize it as an issue.

Some new issues become new issues because we recognize, suddenly, that they are amenable to change. We never thought that health was subject to human control. We either got well or died depending on "God's will." Then we found medicine. We always thought the number of children a woman had was a matter of "God's will," but then we learned about contraception and recognized that this too was subject to human control. What seemed a fixed fact was, in fact, a policy choice—"what is our demographic destiny?" There is a new public policy question that all nations must ask themselves: How many people can live satisfied lives in your nation? In the United States our own birth rates will stabilize at only 340 million, but with immigration we will likely double, and then double again, the size of the nation. Demographic policy is a new and previously unimaginable issue that has thrust itself upon policy matters.

The Transition from Subduing the Earth to Finding and Accepting Our Ecological Niche

Let me briefly outline some additional challenges to sustainability. How do we rally support to prevent problems that seem likely but are uncertain? Even small odds of catastrophe are unacceptable when the stakes are high. Economist Frank Knight urges public policy to distinguish between 'risk' and 'uncertainty.'[11] Risk is when one can calculate the probability of a future outcome, and uncertainty is when one cannot calculate or even guess at a future outcome. When we buy auto and fire insurance, the company can calculate the risk. However, when you encounter a unique event, like the impact of environmental degradation, for which there is no probability estimate, we face an uncertainty. We travel blind the road called uncertainty. We must depend on subjective estimates and hope we have the courage, foresight, and time to correct the problems that surprise us.

Additionally there is the challenge of getting people to sacrifice now for another generation's benefit. What does one generation owe the future? How far into the future? The farther into the future we owe a duty, the harder it is to motivate and the less sense it makes in most cost-benefit thinking. How do we properly evaluate or discount a 200 year future?

The Rio Declaration on Environment and Development states unequivocally "Human beings are at the center of concerns for sustainable development. They are entitled to a healthy and productive life in harmony with nature."[12] This statement speaks to my heart but not to my head. It seems to me that in a limited ecological world we cannot entitle everyone to a healthy and productive life. It seems too much to hope for that our aspirations and the aspirations of our children and grandchildren will fit within ecological limits without substantial disruption. I suggest that we cannot be growth maximizers and ecological realists at the same time.

You cannot maximize all the variables of any equation, and thus it seems improbable that we can realistically ensure that humanity (in the words of the Rio Declaration) "meets the needs of the present without compromising the ability of future generations to meet their own needs." I hope it is true; I doubt that it is.

Look at any American resort community—6,000 square foot second homes, and I venture to state most of their owners still have additional material "aspirations". Humans appear through history to be insatiable creatures, and thus there is an inherent conflict between meeting the "aspirations" of both the upwardly mobile present generation and the presumably upwardly mobile future generations. There is no reasonable limit to "more," "bigger," "faster." The evolutionary traits that allowed humanity to prevail in a cruel Darwinian world now come back to haunt us.

To me, both the Rio Declaration and the Brundtland Commission set utopian but unachievable goals. It attempts to reconcile the irreconcilable, and that poses a substantial challenge to sustainability.

Conclusion

A successful society must constantly evaluate its governance structures. Institutions atrophy and problems change and evolve. Successful governance requires institutions that are equal to the magnitude of the issues they face. Governance structures that worked brilliantly at one time, when faced with one set of problems, fail when confronted with another set of problems. Successful societies have governance structures that meet and manage change.

The current dilemma and challenge is not only that we are faced with unprecedented change; additionally, yesterday's solutions have become today's problems. The transition from growth to sustainability will be one of the great transitions of history. Our economic, religious, social, and political systems all are built on the growth model. It may be beyond the capacity of our institutions to handle this transition without massive disruption. But no generation gets to pick the challenges it is faced with. Sustainability is the current challenge, and it will not go away.

Notes

1 Snow, C.P. 1993. *The Two Cultures*. Cambridge: Cambridge University Press.
2 Berger, P. 1990. *The Capitalistic Spirit: Toward a Religious Ethic of Wealth Creation*. Oakland, CA: Institute for Contemporary Studies.
3 Lippman, W. 1965. *Public Opinion*. New York: The Free Press.
4 Sowell, T. 1987. *A Conflict of Visions*. New York: William Morrow.
5 Morrison, R. 1999. *The Spirit in the Gene*. Ithaca: Cornell University Press.
6 *Ibid.*
7 Shepard, P. 1998. *Coming Home to the Pleistocene*. Washington, D.C.: Island Press/Shearwater Books.
8 Tuchman B. 1981. *Practicing History*. New York: Alfred A. Knopf.
9 Levergood & Breyfogle. 1992. *A Renewed Foundation: A Plan to Rebuild Political Parties and Restore Citizen Democracy in America* (unpublished book manuscript).
10 Bohr, N. Source and date unknown.
11 Frank H. Knight. 1999. What is truth in economics. In *Selected Essays by Frank H. Knight*. Ross B. Emmett, ed.: Chicago: University of Chicago Press.
12 World Commission on Environment and Development. 1987. *Our Common Future*. Oxford: Oxford University Press.

15

Sustainability and Governmental Foresight

Lindsey Grant

"Sustainability" was well defined by the Brundtland Commission:

> "Sustainable development is development that meets the needs of the present without compromising the ability of future generations to meet their own needs."[1]

"Foresight," in its restricted political meaning, is a systematic process of bringing lateral and long-range implications into policy decisions. It is "where the rubber meets the road"—the process of bringing sustainability and the results of academic studies (called "Survival Studies" elsewhere in this book) into the political decision process. It is meant to correct an age-old problem: our decisions are usually the product of tunnel vision, but their most important consequences are frequently the unanticipated lateral impacts.

The "Sustained Growth" Oxymoron

Those concepts have developed as we have come to recognize that human numbers and economic activity pose threats to the systems that support us: air, water, land, resources, and the biosphere itself. There is no "school solution" as to just what is a sustainability issue. Is an economic collapse a sustainability problem? Probably not, because we have developed other decision processes internal to economics, such as Keynesian money flow management. Is war simply a strategic issue, or is it one of sustainability? Certainly either activity becomes a sustainability issue if it threatens our life support systems.

There are dozens of clear sustainability issues, from pollution to biodiversity loss to climate change. But there is fundamental confusion as to whether growth itself is sustainable. The phrase "sustainable growth" appears more often than it should. I would argue that material growth—demographic and/or economic—is unsustainable over long periods. The past century was unique. Perpetual growth is mathematically impossible in a finite space such as the Earth. Assume a starting figure and a growth rate—the rate is more important than the starting level—and I can show when the assumption of continued growth becomes absurd, by any standard you may choose. This is not a theoretical exercise. Current growth reaches the absurd very fast. As the United Nations Statistical Office pointed out in 1992, world population would reach 694 billion people in 2150 if then current fertility and mortality rates were maintained.

Those limits to growth are usually ignored or denied, even by people who profess to be pursuing sustainability. The world has become addicted to growth, and those who profit from it are quick to deny that it must sometime stop.

The Brundtland Commission, after defining sustainability, got off to a wobbly start. Because of its diverse international membership, it could not agree on the need to stop population growth. It foresaw a five- to ten-fold growth in world GNP in fifty years to meet the minimal needs of growing populations. It even concluded that "the international economy must speed up world growth while respecting the environmental constraints." It never addressed how those antithetical goals could be achieved. The WCED report spoke of "sustainable development" rather than "sustainability." Its view of "sustainable development" comes perilously close to the inanity of "sustained growth."

The WCED led to the Rio Conference in 1992 (the U.N. Conference on the Environment and Development, or UNCED), which in turn passed a set of environmental proposals ("Agenda 21"). These led to a follow-up meeting "UNCED+5" in 1997, with another planned for September, 2002. "Sustainable development" has been the key phrase, but the many documents generated by the process have never—literally never—addressed the question: can there be sustainability with continuing growth in demand?[2]

Growth will stop. Will it stop in benign or catastrophic ways? That depends largely on whether or not we recognize what is happening and take steps to deal with it. Sustainability becomes a meaningless slogan unless it deals with the issue of growth. Indeed, I could make the case that the most immediate and important task of Survival Science should be to examine various patterns and rates of growth to learn how long—if at all—they can safely be maintained. We are thus seized of a concept that needs considerable clarification.

The next fundamental question about sustainability is how to put it into practice. Enter foresight.

The Concept of Foresight

The practice of political foresight has been with us for a long time, presumably since some Paleolithic tribal leader contemplated what the consequences might be if he picked a fight with a neighboring tribe or disposed of a rival.

One can hardly expect modern politicians to get much ahead of the current political mood. They may, however, begin to address issues such as the implications of climate change or the consequences of growth if they can be persuaded to ask, in the context of one policy decision after another, "What are the consequences of the proposed policy?" This process is usually referred to as "foresight." It is not simply a concept. It is the institutional machinery to provide a systematic evaluation of the probable consequences of trends or of anticipated actions. It needs:

- an organizational structure to bring different departments and specialists together;
- an ongoing review of environmental, social, and demographic trends and their probable consequences; and most important,
- a means of bringing those conclusions to bear in the decision process.

It is not planning. It does not supersede existing machinery for decision making. Rather, it brings lateral and long term consequences into those decisions. It is, in other words, a process and the people to carry it out, located at the key intersections of decision making in government to require that the question "what are the likely lateral and long term consequences" be asked before decisions are made. It is the creation of an "ombudsman" wired into the White House policy process, whose sole function would be to assure that those issues are brought into decisions.[3]

The Stalled Governmental Foresight Process

In the U.S. Government, there is a long history of efforts to create such a system, even before the term "foresight" came into use. Among those efforts: President Theodore Roosevelt's creation of the U.S. Forest Service and the National Conservation Commission; the 1921 legislation requiring the Executive branch to submit a formal annual budget to Congress; even the traditional and informal cross-agency review and clearance procedures when the interconnections between issues are readily recognized. I will list and briefly describe some of the precedents still relevant today.

President Franklin Roosevelt had an abiding interest in improving the government's capability for coordinated planning. A National Planning Board was established in 1933 to coordinate the planning of public works projects. This board was transformed into a Cabinet-level National Resources Board under the Secretary of the Interior, and by late 1934 it issued a detailed report which claimed to bring "together for the first time in our history, exhaustive studies by highly competent inquirers of land use, water use, minerals, and related public works in their relation to each other and to national planning." The report addressed itself to such problems as "Maladjustments in Land Use and in the Relation of Our Population to Land, and Proposed Lines of Action." It also included an inventory of water resources, a discussion of policies for their use and control, recommendations for a national mineral policy, and a discussion of its international aspects.[4] One wonders how far we have really come since then.

The Board evolved eventually into the National Resources Planning Board, which survived until 1943, when the President dissolved it in the face of unrelenting Congressional opposition.

The Brownlow Report. The modern Executive Office of the President originated in the almost forgotten 1937 Report of the Committee on Management in the Federal Government (the "Brownlow Committee"), created by the President and Congress. Over the years, the bureaucracy had sprawled into over 100 separate organizations, and the President lacked any central staff to coordinate or even to keep track of it. A remarkably concise document, the Report was prepared by a three-man committee chaired by Louis Brownlow. First among its five recommendations was a proposal to provide the President with executive assistants "... probably not exceeding six in number ... to deal with agencies of the Government .... These aides would have no power to make decisions or issue instructions in their own right. They would not be interposed between the President and the heads of his departments. They would not be assistant presidents in any sense. Their function would be, when any matter was presented to the President for action affecting any part of the administrative work of the Government, to assist him in obtaining quickly and without delay all pertinent information possessed by any of the executive departments so as to guide him in making responsible decisions; and then when decisions have been made, to assist him in seeing that every administrative department and every agency is promptly informed. Their effectiveness in assisting the President will, we think, be directly proportional to their ability to discharge their functions with restraint. They would remain in the background, issue no orders, make no decisions, emit no public statements ... ."

That itself was an excellent basis for foresight machinery, but human nature intervened. I can hardly argue that the occupants of the Executive Office of the President have been as self-effacing as Brownlow had hoped, or as dedicated to improving the process rather than advancing their own agendas.

The National Environmental Policy Act of 1969 (NEPA).

Anybody interested in sustainability should reread NEPA. It begins with these eloquent words:

> The purposes of this Act are: [T]o declare a national policy which will encourage a productive and enjoyable harmony between man and his environment; to promote efforts which will prevent or eliminate damage to the environment and biosphere and stimulate the health and welfare of man; to enrich the understanding of the ecological systems and natural resources important to the Nation; ...

> The Congress, recognizing the profound impact of man's activity on the ... natural environment, particularly the profound influences of population growth, high density urbanization,

industrial expansion, resource exploitation, and new and expanding technological advances ... declares that it is the continuing policy of the Federal Government ... to create and maintain conditions under which man and nature can exist in productive harmony, and fulfill the social, economic and other requirements of present and future generations of Americans.

A beautiful statement of the idea of sustainability, before the word itself came into use. The Act then goes on to spell out a process—the Environmental Impact Statement (EIS)—to enable the government to examine proposed governmental actions to see whether they meet those goals.

The Act does not tell the government what it can or cannot do. It was conceived as a "process bill." It tells the government that it must consider the potential environmental consequences of proposed actions, and it provides for public participation. The problem is that the law has been used for limited projects such as interstate highway intersections, but never really applied to major national decisions.

Anybody who wishes to bring long term goals back into the national dialogue could hardly find a better place to start than demanding that NEPA be followed. Ignored though it is, it is still the law of the land.

Concerns about resource depletion led to several post-World War II studies: the "Paley Commission" of 1951, the National Academy of Science's Committee on Natural Resources in 1961, and the National Commission on Materials Policy of 1970.

Population change has been brought into governmental thinking processes only occasionally, usually as a parameter to be reckoned with rather than as a variable itself influenced by governmental decisions. The principal exceptions were three reports.[*]

The Rockefeller Commission (the Commission on Population Growth and the American Future), was created in 1969 by President Nixon and Congress and chaired by John D. Rockefeller III. It concluded in 1972 that further population growth would be detrimental, and it offered suggestions as to how to stop that growth. It was too controversial for the time. The President did not accept it, and it is largely forgotten. It deserves better. The recommendations

---

[*] A detailed look at foresight proposals through 1987 is available in my *Foresight and National Decisions: The Horseman and the Bureaucrat.*

would be a good starting place even now for organizations seeking to make realistic proposals as to how to address the U.S. future.

The Global 2000 Report to President Carter in 1980 had two broad purposes: (1) to present an integrated description of major world trends in resources and the environment and to relate them to population growth; and (2) to evaluate the capability of the U.S. Government to conduct such integrated foresight on an ongoing basis. In a three-volume report, it warned of the dangers posed by current trends. It stated flatly that the government does not have the capability to make integrated cross-sectoral analyses. That was 1980, and it is still true.

Global Future: Time to Act. *The Global 2000 Report* offered no recommendations, but it was followed up in January 1981 at the very close of the Carter administration by a booklet of action proposals from the U.S. Department of State and the Council on Environmental Quality which included eight broad recommendations concerning U.S. population growth. *"The United States should develop a national population policy which addresses the issues of:*

- *Population stabilization*
- *Availability of family planning programs*
- *Rural and urban migration issues*
- *Public education on population concerns*
- *Just, consistent, and workable immigration laws*
- *The role of the private sector—nonprofit, academic, and business*
- *Improved information needs and capacity to analyze impacts of population growth within the United States*
- *Institutional arrangements to ensure continued federal attention to domestic population issues."*

Those last two proposals were expanded into extensive recommendations for foresight machinery, including a proposal for a "Global Population, Resources, and Environmental Analysis Institute, a hybrid public-private institution ... ."

Concern about the environment peaked around 1970 in major legislation dealing with many environmental issues. The legislation was focused on the long term, but in one sense it was deficient as foresight. It tended to set absolute goals (such as zero water pollution), rather than to relate environmental and resource goals to each other or to national goals.

The Global Issues Working Group (GIWG). *The Global 2000 Report* generated widespread public interest. The Reagan administration responded by instructing Chairman Alan Hill of the Council on Environmental Quality (CEQ) to create an interagency group to "identify global environmental and resource issues of national concern, and recommend appropriate government action ... and to improve the U.S. national capability to gather information and to forecast future trends."

The GIWG had a short and dispiriting history. Chairman Hill seemed genuinely enthusiastic, but he was reined in by more powerful players in the White House Cabinet Council. The GIWG was ignored by the powerful, and it became mired in inter-agency disagreements over proposed papers. Launched in 1982, it generated one harmless statement of "Global Environmental Principles" and two position papers for minor international conferences before it simply went dormant about 1985.

The moral is that the pursuit of foresight is doomed unless there are powerful advocates at the top. The CEQ, technically part of the White House, has never had that sort of power. Somehow, the President and his top advisers need to become convinced that this is a process they cannot ignore.

The House Committee on Energy and Commerce Oversight Subcommittee in May 1982 held hearings on various foresight proposals. They were summarized in a Congressional Research Service report, but nothing else happened.

The Critical Trends Assessment Act. During the '80s, Representative and then Senator Gore repeatedly introduced variants of a bill to create a foresight process within the White House (one of which was co-sponsored, surprisingly enough, by his public nemesis, Congressman Newt Gingrich). Three Senate committees held a Joint Hearing on the whole issue of foresight on April 30, 1985. Senator Gore presented his bill, but the committees took no further action. He did not press it. The bill did not really tie foresight into the decision process, but it could still be a starting point for foresight legislation.

The President's Council on Sustainable Development (PCSD). Like President Reagan and the GIWG (above), President Clinton saw that "sustainability" had a constituency. On June 14, 1993, he created the PCSD. Its mandate was equivocal from the start. The President charged it with helping to "grow the economy and preserve the environment ... ," objectives that may be expected to conflict in all but the shortest time frame.

The Council itself was caught in the same conflict. Population and consumption—two of the critical elements of sustainability—were initially not even in its scope. They were introduced, over opposition, at the insistence of Council member and Undersecretary of State Timothy Wirth.

To its credit, in its March 1996 report the Council called for a "Move toward stabilization of U.S. population," (Goal 8) and said that "The United States should have policies and programs that contribute to stabilizing global human population ..." (Principle 12). It remained equivocal, however, as to the broader issue of growth itself. "... [S]ome things must grow—jobs, productivity, wages, capital and savings, profits ... " It did not address the likelihood that more workers with higher productivity are likely to increase environmental stress, even with efforts at amelioration, and that growth itself is at some point unsustainable. It did not, in other words, explore what sustainability really is. It avoided the problem of how to stabilize population by leaving fertility up to "responsible" individual decisions and by avoiding positions on abortion or immigration levels.

The Council was a mixed body with members from the Cabinet, environmentalists, labor leaders, industrialists, and a mix by sex, race and ethnicity. To its credit, that diverse group recognized the need to stop population growth. It was not alone in its inability to face the tough decisions that would be needed to do it. (The PCSD survived in truncated form, through 1998.)

The President was "pleased ... to accept" the report.[†5] This is a small footnote to history. The President's acceptance of the report, casual as it was, constitutes the first explicit acceptance by a President of the proposition that U.S. population should stop growing. But nothing happened to give meaning to that acceptance. Presidents seem to prefer the appearance of action on sustainability to the

---

† Summaries of the Council's work were available at www.whitehouse. gov/pcsd and full texts of its reports at www.sustainable.doe.gov, but these sources seem to have disappeared with the change of administration. For an evaluation, see NPG booknote "Population and the PCSD" (Washington, DC: Negative Population Growth, Inc., April 1996). NPG Forum articles, footnotes and booknotes are available from NPG, and full texts are carried on its website: www.npg.org.

substance—unless, perhaps, they are forced by sustained pressure and an aroused public.

The Science Advisory Board of EPA in January 1995 issued a set of recommendations titled *Beyond the Horizon: Using Foresight to Protect the Environmental Future* (EPA-SAB-EC-95-007). It started with the recommendations that "As much attention should be given to avoiding future environmental problems as to controlling current ones. ... EPA should establish an early-warning system to identify potential future environmental risks." First among the "forces of change" they listed was "The continuing growth in human populations, and the concentration of growing populations in large urban areas ..." Nothing has come of the Advisory Board's recommendation.

There are some lessons to be drawn from that brief catalogue.

Some of the foresight proposals survive; others have perished. There is one profound difference among them. The ones whose legacies have survived, like the Brownlow Committee report (or, to a lesser extent, NEPA) are those that called for the creation of a process. Most of the other studies simply evaluated trends and then disappeared. Such study projects cannot move the nation on vast and complicated issues. Moreover, none of the projects had a specific "peg"—a current issue or pending decision to which they were relevant and which forced their conclusions into policy making. A report published in a vacuum tends to disappear in the face of simple inertia and bureaucratic resistances to change.

In one sense, that catalogue of foresight efforts is a dispiriting one. There has been a curious slackening of interest in foresight mechanisms. The need for them was taken much more seriously in the 1930s, and again in the Nixon era, than it is today. Some of the arrangements have survived. If they have not worked out as planned, it suggests the strength of built-in resistances to objective foresight processes, which I will come to now.

The Resistances

If I have interested the reader in developing a foresight process, let me offer some nuts-and-bolts thoughts as to how to do it. An abstract model of foresight may be of academic interest, but to achieve any real results, any proposal must be constructed to accommodate political and bureaucratic realities. It must be accepted. Once in place, it must work.

In the ways of Congress and of the Executive branch, there are certain patterns of behavior which may initially dispose both branches against foresight proposals. If recognized and properly used, however, these same characteristics may be exploited to gain support for the proposal.

Congress

Theoretically, foresight processes could be introduced in the Executive without a Congressional mandate, but there are solid arguments for going the legislative route. Legislation should reflect foresight. And a statutory process is less likely to be dissolved by a change in the administration than is a solely Executive one.

Dealing with the real Congress. The first obstacle for many proponents of better foresight is their own idealization of how Congress works. One regularly encounters the starry-eyed belief that all one need do is to convince Congress that there is a better way of making decisions, and Congress will forthwith adopt it.

Congress doesn't work that way. Even the most idealistic Congressmen, if they are to remain Congressmen, must respect the cynical old maxim: the politician's first duty is to get elected. They must pay attention to the powerful and the vocal in their own constituencies, and they must trade votes within Congress if they are to pass legislation they want passed. Some Congressmen may even be opposed to proposals that promise to clarify the implications of decisions on which they must vote; it might advertise the cynicism of the vote they plan to cast.

Foresight proponents must build a constituency, either by riding a popular idea (e.g. "efficiency in government") or by enlisting those to whom Congressmen listen, or both.

Building a constituency. Uncommitted legislators will want to see that there is significant support for any proposal, and that their own support would not only be conceptually sound, but likely to improve their political position.

There are groups interested in improved foresight that do not necessarily share the environmentalists' specific concerns. Business is interested in the best available estimates of the geographic and age distribution of populations in the future. Investors and insurers are interested in potential risks. The lumber and building industries will be interested in projections of saw timber resources. Labor and educators share with business an interest in forecasts of the economy and the sectors of growth and stagnation. Educators and local governments are most interested in how large the coming school-age cohorts will be. And so on.

These are all potential allies, to a point. They share the need for good demographic projections, though they may part company when the environmentalist describes the implications of those numbers. Find allies in these groups, and define "foresight" broadly enough to interest them.

Substantive goals vs. procedure. Many legislators know that the present system is deficient in foreseeing tomorrow's critical issues. Not all of them are willing to embrace specific views about demography or the environment. There is an informal principle that process bills—those that establish more or less permanent techniques for dealing with issues—should be divorced from legislation stating a substantive position. Even environmentally-minded legislators may share this belief.

To increase the chance of passage, go for process legislation without mixing it with substantive value judgments. The foresight proponents' purpose will be adequately served if population, resources, and environment are included among the issues to be addressed.

The Committee system. Congress doesn't often override its own committees. Before a bill reaches the floor, it must be reported out of its committee. The way a bill is drafted will influence the decision as to which committees have jurisdiction. Committee chairmen have varying records of support for this sort of legislation.

The history of the Gore Critical Trends Assessment bill is instructive. It kept dying in Rep. Brooks' Committee on Government Operations. Rep. Brooks was not interested in foresight. Rep. Dingell's Committee on Energy and Commerce was much more receptive, and his committee staffers were supportive. Had the language in the bill referred explicitly to the need for better ways to shape energy policy, it might have been concurrently referred to Dingell's committee, with better prospects for a hearing.

The Executive Branch

The Administration's role. Any bill will pass more easily with the administration's support than without it. And after all, the bill must be signed or a veto over-ridden. An administration would be unlikely to make effective use of machinery it didn't want anyway. Don't treat the administration as an adversary.

"Pull, don't push." Sell your proposal as a way of advancing White House goals. Talk with the administration to see whether there is any basis for an approach acceptable to both sides. Population, resource, and environmental arguments are unlikely to generate much interest in the current White House. The argument for more efficient government and a broader constituency seems the most likely point of departure for such a dialogue. Foresight advocates should open a dialogue with key staffers in the individual departments, the OMB, and the White House.

Money. The competition for funding is endless. Proposals to use existing governmental machinery will fare better and gain more allies than proposals to create expensive new machinery.

Manageability. In an increasingly complex world, it gets harder and harder to make wise decisions. The issues must be defined as explicitly as possible, the policy choices kept few enough in number to be graspable, and the potential implications of each choice described in very broad-brush strokes.

Holism. The recognition that everything is somehow connected, but frequently in very arcane and vague ways, is perpetually at war with this need to limit choices and to focus on a few key issues. "The World in XXXX" projections can be a valuable tool—but the cost is likely to be exclusion from current real-world decisions. Use them to buttress policy-specific analyses.

Words. No president has time to read all that is offered him, and they are all faced with bureaucracies that won't believe that simple fact. Don't mistake prolixity for profundity. As any erstwhile staffer knows, the White House staff substitutes a few pages of summary for volumes of briefing papers. That telescoping poses an especial danger to foresight proposals. The staffer may understand his own specialty, but he will probably not understand what the foresight people are saying about lateral implications. If you want to reach the president, keep it brief and dramatic, or somebody else will garble your message.

The "clearance" process. The more broadly defined an idea, the more departments and agencies will have a right to "clear off" on the proposal, and each of them will have its own pet projects which it will try to insert into the proposal. The originator learns to define his project as narrowly as possible to minimize the clearance process, to keep the thing pointed in the direction he wants, and simply to move it through government. This necessity is in eternal conflict with the purpose of foresight to examine lateral implications.

The office managing the foresight process must be interested only in the process itself. It must not be seen as a competitor on policy, or the originating office will try to bypass it.

"Turf." As a principle, any organization will resist any proposal which would deprive it of any responsibility which it now possesses. Proposals to create super-agencies for planning or foresight run head-on into this principle; proposals to strengthen the agencies' own foresight capabilities do not.

Time. Clearances may take weeks or months, particularly if the other agency is reluctant. Major policy studies may take years. Some issues can wait for such delays. Most cannot; and frequently the government is forced to act by external events that do not wait.

Any proposal for improved foresight—which is inescapably a complicating addition—must build time limits into the process. I have watched the Defense Department, which can take months, reply within 24 hours when the White House says, "that's all the time you have."

The "need to know." Leaks are endemic in Washington, as each successive president learns to his anger. The leaks may be generated simply by indiscretion or puffery, but more frequently they are a tactic by opponents of a proposal to give the opposition time to mobilize. Administrations invariably react by restricting knowledge of policies under consideration to those who "need to know."

Nothing could be worse for foresight. Those who decide who "needs to know" themselves do not know what an excluded participant might have been able to tell them about the potential results of proposed action.

The "leakage" problem cannot be completely escaped, but any proposal to widen access to policy documents should show how it will control access to sensitive documents.

"Face." No administration is likely to welcome post-hoc criticism of a decision once made, and such criticism generates an interest in getting rid of the offending voice.

Inject the foresight into the policy process before the decision is made. Don't create a machinery that will tend to second-guess completed decisions.

The urgent vs. the important. Urgent short-term problems regularly take precedence over more important long-term issues as the system tries to keep up with the geometric growth of the issues it must deal with.

Tie foresight to current issues. Show the implications of currently proposed actions. Don't prepare a general briefing about "declining water tables in western aquifers." Show "here's what your proposal for urban water supplies will do to agricultural production, farm income, and exports."

The "black box" syndrome. Many officials are suspicious of complex computerized models. They suspect (rightly) that the results can be manipulated by changing the equations and coefficients in the computer program (the "black box" which they cannot see inside). Global futures projections contain hundreds of such variables and, even so, are too crude an approximation of reality to be used with confidence.

Keep it simple. Begin, perhaps, with projections, "if other factors are held constant." Emphasize that the central element of foresight is not to introduce computer models, but to bring the appropriate human expertise to bear on policy proposals.

Access to the president. Washington is a city of courtiers. The name of the power game is access to the president. There are a few glamorous agencies—State, Defense, Treasury, the CIA—that have such access. Defense would certainly be heard if an EPA proposal about water quality might affect national security; but it is much less certain that EPA would be heard (or even know in advance) if Defense planned to do something that affected water quality.

Most agencies seek access and chafe at their lack of it. They will support any system that improves their access. Broaden the number of agencies with a role in foresight, and you gain backers.

Momentum. Governmental inertia is not altogether bad. It provides continuity and stability. It is also a force more awesome than most people outside of government can imagine. Like the Juggernaut, it tends to crush or brush aside anything that might change its direction. As the CEQ director learned, you need well-placed friends in the White House who are convinced that the system must change if it is to survive.

Mobilizing the Private Sector

The exhortations above are directed at some nameless group that may seek to improve the U.S. government's ability to think ahead.

Such a group is not presently in evidence. In fact, one can legiti-
mately ask whether the government is approachable with such an
idea at this time. If one has such doubts, there is still a way to
advance the use of foresight, and that is to create a private foresight
capability geared to look at pending governmental decisions and to
examine the lateral and long term implications.

Such a foresight group would have some of the characteristics
outlined above. It would draw upon the expertise of academicians
and private think tanks. Aside from policy specific studies, it might
well engage in various collateral activities such as briefing papers,
polls, news conferences, and periodic "State of the Nation" reports.
It might put pressure upon the government to apply the National
Environmental Policy Act of 1969 (NEPA) to major governmental
decisions. To gain political clout, it might organize "Coalitions for
Sustainability" with other interested groups. If the opportunity presented, it could be the lead group in pressing for better governmental foresight. An opportunity may arise to revive and improve
proposals such as the Critical Trends Assessment Act.

It might be called a "Sustainability Project," hooking into the
environmental community's current enthusiasm for sustainability,
or perhaps a "Foresight Project," to bring the need for better foresight back into the national debate.

The project should be focused on the United States. Anodyne
generalizations about other people's problems are a popular way to
avoid addressing our own.

It would be an action group dedicated to the proposition of sustainability, not a debating club. There is no shortage of debates
elsewhere. It will need to hammer away at its agenda: what are the
implications of what the nation is doing? Simply stating the need for
sustainability is not enough.

"Sustainability," properly used, is an exciting idea, and foresight is a
way to give it substance. It can be the vehicle for moving the United
States away from unthinking reliance on laissez-faire toward a systematic way of recognizing our obligations to future generations,
establishing our social goals, understanding how they relate to each
other, and successfully pursuing them.

Notes

1 Brundtland Commission. 1987. *Our Common Future, Report of the World Commission on Environment and Development.*
2 *Sustainability, Part III. Climate, Population and UNCED+5.* NPG FORUM paper. Washington, DC: Negative Population Growth, Inc., 10–1997. Online at www.npg.org.
3 Grant, L. 1988. *Foresight and National Decisions: The Horseman and the Bureaucrat,* Lanham, MD: University Press of America.
4 *Ibid.*
5 *Ibid.*

16

Educating World Leaders

Christopher Williams

Introduction

Global leadership is the pivotal point for appropriate policies and action to ensure human survival. But how do leaders maintain their knowledge and awareness in a fast-changing world of increasingly complex problems? How do we create a "learning leadership" that will assure the survival of humanity?

Despite the relevance of global leadership for human survival, the academic world has shown little concern with this aspect of "up system" research. There has been an interest in how power groups are created and reproduced, stemming from the work of the Italian "elite theorists" Pareto and Mosca early in the 20th century.[1] But this remains in the realm of social theory and does not contribute much to the practical development of contemporary leadership. Management theories of leadership are practical, but apply essentially to commercial interests, which are often antithetical to the interests of humanity and to our survival. Kennedy Graham concludes:

> "Research into leadership issues is a young and still rather shapeless discipline. While leaders and leadership may provide the

stuff of bar-room wisdom and talk-show humour, it is an elusive subject from which to glean analytical insight and prescriptive value at a level approaching normal academic standards. While some literature exists offering unfalsifiable theories about leadership behaviour and personality, there is a dearth of primary empirical information about leaders, the philosophical prisms through which they perceive reality, and the principles by which they conduct themselves. This is especially this case with leadership on global issues."[2]

Leadership is also under-theorized in many areas related to the survival debate, for example national development,[3] social movements,[4] and the education of leaders.

The United Nations University Leadership Academy (UNULA) in Amman, Jordan is one of the new organizations trying to address this situation. This discussion reflects ongoing research and course development at the UNULA, and examines two key questions. Why should we be concerned about the relationships between education, leadership, and survival? How do potential and serving leaders acquire the necessary attributes for them to recognize and address the threats and challenges to our survival in the contemporary world? But first, what exactly is "global leadership" in relation to survival?

Global Leadership and Survival

A mention of global leadership inevitably demands definitions of "global" and "leadership," which often then consume a large proportion of the debate with no workable conclusions. For the purposes of this discussion, a compromise between conflicting claims assumes that "global" can embrace circumstances that reflect human universals, an extended national or regional interest, and/or the condition of the global commons.[5]

Similarly, "leadership" has as many definitions as texts. These are constantly evolving, and few embrace a global dimension.[6] The basic concept of leadership within the standard Oxford Dictionary definition is "guiding". This is reflected in other languages. The Chinese character for leader means a "pointing leading person". The Hausa word *Shugaba* implies "the one in front."[7] But not all traditions recognize this role. For example, the Maori understanding of chieftainship is more akin to "trusteeship".[8] In this discussion, global leadership is taken to describe, "guidance through act, influence, or inducement, including financial arrangement, by individuals, collective leaderships, organisations, or States, which has actual or possible consequences affecting a human universal, an extended national or regional interest, or a common global and planetary interest."

Leadership may appear in many forms and derive from many sources, but all are likely to guide through act, influence, or inducement.[9]

The characteristics of a globalising world are increasingly seen as constructing a mutual interest beyond that of nation-states. This proposes that national leaders should accept that the national interest could be superseded by a greater world interest in some circumstances, which raises the question of the criteria determining that interest. Within the quasi-legal concept, the "planetary interest," Kennedy Graham provides a pragmatic definition on two levels analogous to the accepted conventions delineating the national interest:

> "The 'vital planetary interest'—the survival and viability of humanity contingent on the maintenance of the physical integrity of Earth and the protection of its ecological systems and biosphere from major anthropogenic change ... the continuation of the planet in its state of natural equilibrium.

> "The 'normative planetary interest'—the universal improvement in the human condition in terms of meeting basic human needs of each individual and observance of the fundamental human rights of each..."[10]

It is therefore meaningful to talk of a common global or planetary interest, in relation to human survival.

The main concern of this discussion is senior people at national and international levels. But, in the view of Graham, the global dimension extends beyond high profile individuals, and entails "... leadership exhibited by men and women anywhere in the world, that serves the 'planetary interest' of human unity, inter-civilization harmony, global security and sustainability, and environmental integrity. It can be exhibited not only by leaders of the most senior international positions. It may be exhibited by middle-level managers...[or] the head of the most humble household who teaches her children the art and virtues of domestic recycling."[11] Although the focus is individuals with political power, many of the lessons are equally relevant to, and can be gleaned from, all levels of society.

The central implication of "survival" is basic survival—the long-term continuance of humanity and the planet. The significant problems for research, as defined by John Herz, are "questions of how human survival can be ensured in the face of the threats of extinction with which the entire human race is now confronted."[12] But other forms of continuance are also relevant. In relation to global security, Gwyn Prins emphasizes *positive* survival—"a condition above bare existence" marked by the things that make us human and distinct from other forms of life.[13] This entails continuance of knowledge, social groups, social systems, families, and what UNESCO now calls "indeterminate heritage"—traditions, philosophy, systems of calculation, poetry.

From the perspective of global leadership, assuring basic survival and ensuring positive survival are not categorically different, as they might at first seem. It is likely that a despot who is willing to promote regional genocide would also readily employ weapons of mass destruction and threaten humanity. There seems little doubt that Hitler would have used the A-bomb, had it been available to him. The U.S. advocate of nuclear deterrence, Joseph Nye, Jr., unashamedly maintained that the total destruction of the human race was preferable to an existence that did not preserve positive survival in the form of the American way of life.[14] We need to understand the mindset of powerful people who think like this.

Other new security threats similarly embody a link between basic and positive survival. Leaders who permit "small scale," "local" trials involving genetic modification are making a decision with global implications, although they rarely admit it. Similarly, the failure to address local outbreaks of diseases such as BSE/CJD ("mad cow disease") will have consequences for human survival well beyond a single country. Suicide killings, such as those on 11th September in New York, evidence that the killing of self, adversaries, and innocents can be justified to achieve a particular ideological goal. If pursued to its logical end, that ideology leaves everyone dead. What is the education that has led to that view? Some leaders deter the use of contraceptives and jeopardize humanity through overpopulation or HIV/AIDS. They are also promoting policy on a local level that, if promoted globally, could threaten the survival of humanity. Although seemingly diverse, many challenges at the level of positive survival have implications for the basic survival of humanity, not least for the positive survival of educational traditions.

There is another, more pragmatic, reason for considering "positive survival" in relation to leadership. The total obliteration of humanity within a foreseeable time frame would require the occurrence of events that appear improbable. Our complete demise can only happen in one of four ways:

1. the simultaneous death of all humanity
2. degradation of life support systems to a degree that brings about the death of all humanity
3. a universal disease, toxin, or other environmental impact that kills before reproductive age
4. a universal disease, toxin, or other environmental impact that prevents human reproduction

Impacts that do not fulfill one of these criteria might devastate humanity, but not eliminate us. As I have argued previously, the decline of human intelligence caused by adverse environmental change is one possibility. And paradoxically, less intelligent human beings would not merely survive. They would probably survive longer, because they would not do "intelligent" things such as building nuclear weapons or modifying their DNA. But in such a society, a visit to the dentist would not be much fun.[15]

As the legends of creation remind us, provided one man and one woman with reproductive capability are alive, the human species could continue in some form. Political leaders are aware of that, and while paying lip service to the threat of extinction, it is unlikely that they will take it seriously. Despite the survival-oriented *raison d'etre* of the United Nations, research carried out by the UNULA in 2001 finds that even the leaders of the international organizations do not overtly concern themselves with fundamental human survival. The world problems as seen by Juan Somavia, of the International Labour Organization (ILO), are "poverty" "social exclusion" and the need for "decent work." At the World Trade Organization (WTO), Mike Moore identifies, in the last century, European tribalism and the "twin terrors" of Marxism and Fascism as the main threats. He sees "those who are locked out" of the information revolution, and the possible failure of the multilateral system, as challenges in the future. Unsurprisingly, from the perspective of the World Health Organization (WHO), Gro Harlem Brundtland considers "the health of people" to be very significant. But, as one would hope of the progenitor of "sustainable development", she also expresses a link between basic survival and positive survival. "Healthy people—healthy planet", she concludes.[16]

Like it or not, contemporary political leaders are going to win more votes through preserving the American way of life than through preserving humanity. And while our only form of world governance is international rather than global, the formal views of the international leaders are likely (and not wrongly) to reflect those of their paymasters.

Why the Concern?

The significance of the relationship between leadership and education goes beyond the abilities and conduct of leaders themselves, because these same people exert considerable influence over the education of everyone else. A contemporary South Korean writer, Yun-Joo Lee, puts the point succinctly:

"Although leaders are a 'minority,' arguably they are more important [than the populace] to examine because they are producers as well as consumers of the services that build human capital. Leaders are a product of their education, either accepting or rejecting the values being taught to them and then choosing to replicate or not the system they experienced."[17]

Any national education system is to some extent a reflection of the ideology of those in power. For centuries leaders have used and abused education systems to achieve their ends, and they continue to do so. Education can extend loyalty to a leader into loyalty to the *ideas* of a leader, and that is attractive to saints and depots alike. Notoriously, education was used to promote Apartheid in South Africa and U.S.A, social hatred in Nazi Germany, and harmful nationalism in pre-war Japan.

This problem is not restricted to past eras. In 1998, the Taliban closed private girls' schools, ruled that other schools could not instruct girls older than eight, and that they could only teach the Koran. In Russia, the supreme mufti Ravil Gaynutdin has expressed concern about Islamic colleges that are promulgating the teachings of Wahabism. This 18th century movement calls on its followers to treat all other Muslims as enemies.[18] Current textbooks teach that anyone who is not a Wahabi should be killed.[19] That ideology amounts to proposing the extermination of most of humanity.

Islamic fundamentalism is not the only source of abuse. In some southern states of the U.S., creationists impede knowledge about evolution and Charles Darwin. In 1929 the Arkansas and Mississippi education authorities banned the teaching of evolution, and the courts have argued over this ever since. The Kansas Board of Education inserted "creationism" into the school standards on science in 1999,[20] which was later reversed by the courts, and then Ohio tried introducing the idea of "intelligent design." The significance of this goes beyond a curriculum nicety. Through understanding evolution, students learn that human beings have survived and evolved for millions of years in relation to a comparatively stable "natural" evolutionary environment. This understanding underpins a questioning of whether or not we will continue to survive in the context of adverse human-made environmental change which radically alters that environment.

Commercial interests are also a concern. In Nigeria, Shell runs a significant number of schools, and Mobil has similar influence in the Amazon region. Are these schools likely to present an unbiased view about global warming? Would a nation achieve a balanced democratic debate about genetically modified organisms if many of its population had been to schools or colleges run by Monsanto? It seems unlikely in the light of Monsanto's efforts to manipulate not only genes, but also the academic world, through articles in medical journals as part of "a planned series of outreach efforts to physicians."[21]

In times of political turmoil, universities often provide the only effective check on wayward governments, and for centuries this has engendered politically motivated interference. Academics face personal danger, and military or political security is not the only rationalization for interference. In Turkey, an expert on cyanide received an 18-month jail sentence in July 2000 because his opinions about environmental and occupational safety could affect the gold mining industry.[22] Outspoken academics may be one of the last lines of defense for human survival. Even presenting data that appears inconvenient to the biotech industry can culminate in the premature ending of an academic career, as biologist Arthur Puzstai found when he suggested in Britain that genetically modified potatoes might affect the immune system of rats.[23]

Leaders do not only influence the education of others through their control of education systems and academia. They also do it directly through speeches and public information, and they do not always do it honestly or wisely in relation to human survival. Thabo Mbeke's idiosyncratic claims about the absence of a link between HIV and AIDS are an obvious example of the latter, but misguided views are not the only problem. There is an intrinsic unwillingness to educate the public about global problems of any type. Naïve cheerfulness is more likely to win votes.

Politicians usually only ride on the threat of danger retrospectively. U.S. President Bush showed little interest in global matters before 11th September.

The symbiosis between leadership and education is not surprising. It is even reflected in etymology: "education"—*educare, educere*—to educe, to lead forth. And "pedagogue" derives from the Greek, *agein,* to lead. If this relationship were more clearly part of public awareness, it might be asked why, although the state has legitimate control of both the means of violence *and* education, and both can be used to threaten our survival, only the former is regulated at an international level. Governments seem more ready to accept international intervention and norms about what happens in their armies than in their classrooms. Perhaps they are all too aware that the pen really is mightier than the sword.

How Do Leaders Learn?

The most relevant aspect of leadership and education is, of course, the specific education of leaders. Throughout the world, beyond private tuition, this has been evident in four forms:

- Elite-creating national systems of education
- Specialist university courses and departments
- In-house pre- and in-service training
- Mid-career tuition at specialist institutions

Again, history provides insights into how survival may be jeopardized if these systems are inadequate or abused.

In many countries the whole of the national system of education has been designed to create elite cadres of politicians and civil servants. France is a common example, which has global relevance because it has been replicated in many of the Francophone countries. The "Oxbridge" system in Britain is similar but less formalized, and the idea also remains evident throughout Britain's former colonies, including the U.S. with its "Ivy League" system. In Japan, the Todai system, with Tokyo University at its pinnacle, continues and has similarly had international influence within the former occupied countries, such as South Korea.

These systems are increasingly seen as ineffective. Selection processes are often inefficient and non-meritocratic. Tuition is inappropriate, and does little to engender an informed global leadership. Students are still required to study academic subjects that are not relevant to their future careers in great and unnecessary depth, and there is rarely any conceptual linking or overview of subjects. Discipline-based intelligence is overvalued, yet studies of senior leaders find that this mental capacity is secondary in its significance in comparison with skills such as creativity.[24] Paradoxically, the traditional systems that lay claim to selecting and educating the future world leaders at the undergraduate level hardly ever have the study of "leadership" on their curriculum, except within management courses or as an aspect of history. They usually do not provide the right input or ethos for those who need to acquire generalist skills, to engage in global "joined-up thinking," and to make increasingly "big" decisions in the context of risk and uncertain knowledge. And they rarely encourage or facilitate ongoing learning.

The result is personified in British Prime Minister Tony Blair, whose Oxford education gave him an excellent grasp of law. But on his own admission, he wishes in retrospect that he had studied international relations—and that he could use a word processor efficiently. In the international sphere, Director General of the WTO, Mike Moore, would like to have learned languages, and states with great honesty, "I wish I had a degree in economics."[25]

Elite-creating systems are open to abuse. Nazi Germany provided an exemplary model of leadership education, at least in terms of efficiency. In Hitler's Germany, the "new nobility" was to be created, from those "of hereditary Nordic substance as has been proved by actual deeds." A decree in 1937 established the *Adolf-Hitler-Schules*. A quarter of these graduates were to go to the "Castles of an order" *(Ordensburg)* for another four years of leadership training. A further selection was to be made to create "leaders of leaders" *(Junkers)*. The Fuhrer was clear that, "[T]he best form of State is one which with the most natural certainty raises the best brains of the community to influence and leadership."[26]

This form of abuse is not just historical. Some of Pakistan's *madrassahs* provide a contemporary example. Before the events of September 11th, Sami-ul Haq, the head of the Darul Uloom Haqqani, one of the top Islamic seminaries, stated that he was proud to have produced some of the "top leaders of Afghanistan's repressive Taliban militia," including interior minister Khairule Khairkhwa, and the head of the feared religious police, Qalam Uddin.[27] Reports from Peshawar, Pakistan told how young boys are taught to "Pray for the destruction of the West, pray for the West to be divided into pieces so that it can be attacked, just as the West has done to Islam."[28]

Any system of education that aims to create and shape elite cadres has the potential to promote an ideology that may threaten human well-being and survival. We should perhaps look more closely at the ideologies of limitless consumption, environmental sacrifice, and "growth" which pertain in the systems in North America, Europe, and East Asia. The abuse of education systems by self-declared despots is easy to identify, and therefore to eventually redress. The threat to survival arising from the ideology of unbridled progress is latent and insidious, and so far has attracted little comment.

The second form of leadership education is through specialist courses and departments, usually at postgraduate level. The U.S. universities have many such courses which are, in general, commercially oriented. Beyond this "leadership" can be part of training for service within religious organizations or other professions, for example "educational leadership," which essentially means upgrading teachers. Courses such as that at Westminster University, London, offer tuition for young diplomats and other international civil servants, within which language learning is central. Increasingly, political leaders have experienced this sort of education; for example, the environmentally-oriented president of Costa Rica, J.M. Figueres, attained a Masters' in public administration at the John F. Kennedy School at Harvard University.

At postgraduate level, the Jepson School of Leadership Studies, University of Richmond (U.S.A), is virtually unique in providing an undergraduate leadership course. This is an interesting innovation, but it begs the question as to whether potential leaders can self-select, and/or be selected, in their late teens.

Within the third form, pre- and in-service training selection is far less problematic. In the commercial sector, management training is now common. But the most relevant aspect is civil service training. These institutions have their antecedence in the Chinese Imperial College (guozijian), built in 1306 by the grandson of Kublai Khan. The Chinese demonstrated that national elites could be created through a meritocratic system, and that this had benefits for governance. The tradition continues at the National School of Administration in Beijing. It was not until the end of the 19th century, prompted by the Northcote-Trevelyan Report in 1853, that the British system of civil service promotion and training adopted a more meritocratic form. The Civil Service College then provided a model for many similar institutions around the world. At an international level, the United Nations Staff College in Turin trains U.N. officials.

In-service training addresses the need for ongoing education, it can be melded into daily practice, and time-efficient delivery addresses the problem that senior people are busy people. But inevitably in-house training suffers from introspection. It is unlikely to encourage fundamental questioning of governmental practice—for example, that civil servant pensions are invested in the genetic engineering industry.

The fourth, and more recent, approach is short courses through specialist organizations. These usually target mid-career professionals, can provide a "second chance" training for people who did not have access to the elite systems, and are often more relevant to human survival. The intent is to provide a brief period of skill enhancement and global awareness just before individuals are promoted to senior positions. Lead International, for example, provides courses promoting sustainable development for an international clientele from diverse professional backgrounds. In the U.K., Forum for the Future employs a similar strategy at a national level. Courses usually embody two aspects: a particular theme, such as water management, and core leadership skills.

The "how" of learning—the pedagogical style—in these courses is very far from that of a traditional discipline-based university. At the UNULA, the instigator of the Academy, former Prime Minister Abdul Salam Al-Majali, established the approach. He uses the phrase, "Expose, don't impose." Potential leaders, he maintains, are not temperamentally suited to formal lectures, but learn experientially, principally through exposure to more experienced leaders.

The difficulty faced by these specialist organizations is their small scale and limited resources, which means that selecting appropriate candidates is crucial. If fees or other costs are prohibitive, or the routes of selection are limited, they can unwittingly replicate the unmeritocratic selectivity of the old elite-creating education systems.

Creating A Learning Leadership-Cybernetics

In addition to the difficulty of selecting the right people, these four forms of leadership education embody another problem: they rarely influence serving political leaders. To be realistic, this will probably always be the case. So how do we engender a learning leadership? Leaders might not be persuaded to take time-out themselves, to upgrade their abilities, but they might be persuaded to create learning-oriented institutions around them—a learning leadership, if not a learning leader.

Serving leaders would claim that they learn "on the job" from one another, from their advisors, and from the population they serve, so we must build on that. This reflects two relational aspects of leadership theory: peer interaction and *reciprocal* relationships,[29] and *transactional* relationships—mutual give-and-take—between leaders and followers.[30] But this is not so straightforward on a global scale. Relevant peers are scattered across the world, and it is hard to envisage a truly transactional relationship between leaders and followers on a global scale.

A third relational interaction therefore becomes significant—*cybernetic* learning—which complements the other two forms. In any biological system, even on a global scale, cybernetic systems provide the feedback that is vital for any form of control to be effective. The term cybernetic is very apposite in relation to leadership—it stems from the Greek *kybernetes,* which means "steersman." Theories of cybernetics have been utilised to explain how commercial leaders learn "on the job," but this idea has not been extended to a global scale.[31]

Norbert Wiener foresaw the significance of cybernetics in the 1950s, in *The Human Use of Human Beings: Cybernetics and Society* (1954). In the view of Jeremy Rifkin, "Weiner came to view cybernetics as both a unifying theory and a methodological tool for reorganising the entire world." Rifkin continues, "Virtually every activity of importance in today's society is being brought under the control of cybernetic principles."[32] There are four aspects that relate to global leadership: the *behaviour* of humanity, the *condition* of humanity, *planetary information,* and *global ethics.*

First, global leadership must attend to the *behaviour* of the global population—indicators reflecting the choices and decisions we make. In the 1960s and 1970s international aid agencies realized that the day-to-day behaviour of nomadic Africans, not high-tech weather projections, was one of the most reliable indicators of impending famine. If the nomads were selling cattle and becoming less willing to make long risky journeys, famine was likely. Scanning local newspapers for conflict-related keywords is being used as an indicator of unrest in Central Asia. Similarly on a global scale, behaviour such as migration and asylum seeking, or a decline in birth rate as in Japan and Russia, provides indicators of planetary disruption or the changing dynamics of globalisation and modernity.

Second, the *condition* or "health" of humanity is indicated by changes such as an increasing birth/survival rate since the 1950s, a possible decline in fertility rate due to chemical impacts, an increase in mental health problems, or pandemics such as HIV/AIDS. A global leader must become a global epidemiologist, seeking out and analyzing effects in the whole population that are hardly evident in individuals or single nations, and noticing effects in small populations which may have global implications, such as Ebola. But this should also embrace a global epidemiology of success—how certain policies and strategies improve the human condition.

The third aspect is that leadership should attend to *planetary information*, for example in the form of changes to climate, biodiversity, radiation levels, oxygen levels, or water availability and quality. Agreements such as that at Kyoto may seem inadequate, but they represent a completely new era in which global leaders attend to planetary information.[33] The term "sustainable" stems from the Greek verb *prosecho*—to take care and *to take notice*. The case of ozone depletion also provides a reminder that "attend" should mean "take notice." When data indicating the thinning of the ozone layer was first recorded by scientists, it was ignored on the belief that it was a mistake.

The idea that the planetary and ecological systems, including human life, can be viewed as a global information system has been developing for three decades. In the 1970s James Lovelock's Gaia Hypothesis introduced the perception of the Earth as a single system.[34] Inspired by the mapping of the human genome, recent innovations propose that all the genes in an ecosystem could be considered as a single genome, and these "eco-genomes" could then be compared to discover rules for how ecosystems are put together.[35] Howard Bloom sees planetary life as a representing an evolutionary "global brain,"[36] and Alison Jolly foresees an increasingly networked information system that will eventually make the planet function as a single "superorganism."[37]

New surveillance technology—for example, remote sensing and Geographical Information Systems (GIS)—increasingly provide the means for leaderships to observe their planet in a way that has never before been possible.[38] Websites such as *Globio* and the United Nations Environment Programme's UNEP/GRID provide a wealth of planetary information about environmental change, yet the potential for improving human well-being is not properly exploited.[39] The likely impacts of the flooding in Mozambique in 2000 were predicted by satellite within minutes, but the information was slow to reach the people who needed it, so remedial action was delayed. The U.K. non-government organisation (NGO) Aid-to-Aid, which makes satellite information available to other NGOs engaged in humanitarian assistance, addresses this problem. But proposals to create an inexpensive ring of remote sensing devises, which could help predict and assess the impacts of natural and human-made disasters, has been languishing with the British civil service for three years.[40] Hopefully, the United Nations Development Programme (UNDP) Millennium Ecosystem Assessment, which links the efforts of 1500 scientists to assess 16,000 satellite images, will greatly improve this information deficit.

Well before these global visions and technological advances, information biologist Gregory Bateson defined information very simply as "news of difference."[41] And he continues with an obvious point that provides an important warning: "all perception of difference is limited by threshold. Differences too slowly presented are not perceivable." This raises a crucial point. Humans are not adapted to notice the "creeping disasters" such as climate change and ozone depletion. The perception of global risk must be intellectual, not sensory,[42] and that is a concern for education, particularly the public information put out by leaders.[43] This requires new forms of collaborative leadership between scientists, technologists, educators and politicians.

But whatever the system, strategy, or pedagogy, there remains the standard dilemma in any discussion of leadership education—it is not adequate to develop technical proficiency without a moral dimension. Technical proficiency alone is as likely to produce a Gandhi as a Hitler. It is also clear that the survival of humanity is inextricably linked to "positive survival" of education, and that is largely a moral question. The fourth aspect is that those with power must also be constantly aware of evolving *global ethics*, particularly in relation to leadership.

There are currently no international standards that specifically address the moral conduct of leaders, simply a proposal from the UNULA. Within this code, of relevance to this discussion, leaders should recognize a moral duty to maintain their own ongoing learning: *"Leaders and their agents shall seek out and consider properly all information and evidence relevant to their duties from all possible sources, particularly in relation to hatred, harm, or hazard, including information from the global population and planetary and ecological systems."* [44] Any set of ethical standards for leadership also proposes a relevant curriculum for the education of leaders. For example, if it is agreed that leaders should respect international law, it follows that they should know the basics of international law. If they should uphold the "global and planetary interest," they should know what that is and the precedents upon which the idea is based.

Some of the centuries-old Indian prohibitions on leadership from the traditions of Kautilya make the point about leadership ethics very clearly: "keep away from another person's wife," "avoid daydreaming, capriciousness, falsehood, extravagance," and "avoid association with harmful persons."[45] There are certain of our modern leaders who might have enhanced their personal potential for survival, had they learned these lessons at some point. And that indicates the symbiosis that must be formed if we are to achieve the broader goals of this discussion. We need to create a conspicuous coincidence of interests between the personal survival of leaders and the survival of humanity, through democratic and more innovative forms of global leadership accountability. In that context, the current dilemmas of leadership education would probably become self-solving.

These four aspects of cybernetic learning embody nothing new. But at present they are rarely viewed holistically as a core aspect of the duties of leadership. We need leaders who can steer humanity through intelligent responses to global information, not through out-dated beliefs or self-serving ideology. A "learning leadership" is one of the most important manifestations of applied survival research.

Notes

1 Meisal, J.H., ed. 1965. *Pareto and Mosca.* Englewood Cliffs, N.J.: Prentice-Hall.

2 Graham, K. 1999 *The Planetary Interest.* London: UCL Press.

3   Lee. Y-J. 2002. *Leadership Education and Experience, and Development: A Comparative Study of S. Korea and Egypt.* Unpublished PhD in-progress, School of Oriental and African Studies, London.
4   Barker, C., A. Johnson, & M. Lavalette. 2001 *Leadership and Social Movements.* Manchester: Manchester University Press.
5   Williams, C. 2001. *Leaders of Integrity: Ethics and A Code for Global Leadership.* Amman: U.N. University Leadership Academy.
6   Ciulla, J. B., ed. 1998. *Ethics: The Heart of Leadership.* Westport: Praeger.
7   Abolarin, A. 1997. The cultural language of leadership. *LEAD newsletter* 7: 5.
8   Kawharu, H. 2001. *Notes on The treaty of waitangi.* At http://www.govt.nz/aboutnz/treaty.php3
9   Williams, C. 2001. *Leaders of Integrity: Ethics and A Code for Global Leadership.* Amman: U.N. University Leadership Academy.
10 Graham, K. 1999. *The Planetary Interest.* London: UCL Press.
11 Graham, K. 2002. *Studies in Vision and Management: Interviews with U.N. Leaders.* Unpublished paper, Amman: U.N. University Leadership Academy.
12 Hertz, J. H. 2001. *Some Observations on Engaging in "Survival Research."* Unpublished paper. Scarsdale, NY.
13 Prins, G. 1993. *Threats Without Enemies.* London: Earthscan.
14 Nye, J. 1986. *Nuclear Ethic.* New York: At the Free Press.
15 Williams, C. 1997. *Terminus Brain: The Environmental Threat to Human Intelligence.* London: Earthscan.
16 Graham, K. & C. Williams, eds. 2002. Global Leaders Interviews Series:
    No. 1 Through the Eyes of People—An Interview with Juan Somavia.
    No. 2 Freedom Is a Universal Value—An Interview with Mike Moore.
    No. 3 Healthy People—Healthy Planet—An Interview with Gro Harlem Brundtland.
    Amman: U.N. University Leadership Academy.
17 Lee. Y-J. 2002. *Leadership Education and Experience, and Development: A Comparative Study of S. Korea and Egypt.* Unpublished PhD in-progress, School of Oriental and African Studies, London.
18 Lane, E.W. 1904. *The Modern Egyptians.* London: Dent.
19 Rich, V. Terrorist colleges rife, says leader. *Times Higher Education.* Supplement. 21 July, 2000.
20 MacKenzie, D. 2000. Unnatural selection. *New Scientist* 22 April, 35.
21 GeneWatch 2000. Monsanto's "desperate" propaganda campaign research global proportions. *GeneWatch* Press Release, 6 September, 2002. At www.genewatch.org/Press%Releases/pr15.htm
22 Jones, D. Cyanide expert jailed for meeting. *Times Higher Educational Supplement* (UK), 7 July, 2000.
23 BBC. 1999. GM research row moves to internet. Available at http://news.bbc.co.uk/hi/english/sci/tech/newsid_368000/368089.stm

24 Rusmore, J.T. 1984. *Executive Performance and Intellectual Ability in Organizational Levels.* San Jose, CA: San Jose State University, Advanced Human Systems Institution.

25 Graham, K. & C. Williams, Eds. 2002. Global Leaders Interviews Series:
No. 1 Through the Eyes of People—An Interview with Juan Somavia.
No. 2 Freedom Is a Universal Value—An Interview with Mike Moore.
No. 3 Healthy People—Healthy Planet—An Interview with Gro Harlem Brundtland.
Amman: U.N. University Leadership Academy.

26 Kneller, G.F. 1941 *The Educational Philosophy of National Socialism.* New Haven, CT: Yale University Press.

27 McCarthy, R. 2000. Pakistan's clerics balk at school reforms. *The Guardian,* 13 September.

28 Carlton, G. 2001 In the frame: Peshawar, Pakistan. *Independent Magazine* (UK), 29 September.

29 Williams, C. 2001. *Leaders of Integrity: Ethics and A Code for Global Leadership.* Amman: U.N. University Leadership Academy.

30 Bass, B. M. 1990. *Bass & Stogdill's Handbook of Leadership: Theory, Research, & Managerial Applications.* New York: Free Press.

31 Hampden-Turner, C. 1992. Standing at the helm. In: *Frontiers of Leadership.* M. Syrett & C. Hogg, eds. Oxford: Blackwell.

32 Rifkin, J. 1998. *The Biotech Century.* New York: Tarcher/Putnam.

33 Kyoto GIS. 2001. At http://maps.grida.no/kyoto/

34 Lovelock, J. 1979. *Gaia: A New Look At Life On Earth.* Oxford: Oxford University Press.

35 Leslie, M. 2001. Tales of the sea. *New Scientist* 2275: 32–35.

36 Bloom, H. 2000. *Global Brain: The Evolution of The Mass Mind.* London: John Wiley.

37 Jolly, A. 1999. *Lucy's Legacy: Sex and Intelligence In Human Evolution.* Cambridge, MA.: Harvard University Press.

38 Harrison, P. & F. Pearce. 2001. *AAAS Atlas of Population and Environment.* Berkeley, California: University of California Press.

39 Globio 2002. At http://www.grida.no/db/gis/prod/html/toc.htm/

40 Graham-Rowe, D. 2001 Saving planet Earth. *New Scientist,* 23 June.

41 Bateson, G. 1979. *Mind and Nature.* New York: Ballentine Books.

42 Seidel, P. 1999. *Invisible Walls: Why We Ignore The Damage We Inflict On The Planet ... and Ourselves.* New York: Prometheus Books.

43 Williams, C. 2002. New security risks and public educating: the relevance of recent evolutionary brain science. *Journal of Risk Research* 4(3).

44 Williams, C. 2001. *Leaders of Integrity: Ethics and A Code for Global Leadership.* Amman: U.N. University Leadership Academy.

45 Waslekar, S. 2000. *Dharma Rajya: Path-Breaking Reforms for India's Governance.* Delhi: Konark Publishers.

17

A Primer of Civilization

James Lovelock with Peter Seidel

We lead our lives as if every day will be like the one before it. Every so often, something happens somewhere that tells us we cannot count on that. A hurricane, an earthquake, a volcanic eruption, or a terrorist act occurs, killing thousands of people. We then scramble to take precautions to prevent these particular events from happening again, to be ready if they do, and to minimize the damage. Often the likelihood of such events could have been foreseen, but we choose not to notice. We would rather lock the barn door after the horse has been stolen.

When we are thirsty, we walk to the sink, turn the faucet, and water comes out. If it is cold, an electric current turns on our furnace which burns fuel from a source we do not see. Or we take a stick of wood from a pile and put it into the fire. We take all this for granted. But what if the water does not flow, the furnace does not ignite, or there is no wood to put on the fire? These conveniences depend on precarious complex systems—organic, technical, or economic. Are we prepared should they not work … or will we wait until they do not?

But there are bigger threats ahead. We have not eliminated the possibility of an all-out nuclear war, global warming may be headed our way, the world's population is growing out of control, or an asteroid could hit our planet, causing major havoc. We have no guarantee that our society will not simply break down under stress as all others in the past have. Would civilization be able to survive and recover?

We have confidence in our science-based civilization and believe it is here to stay. But we are confusing the lifespan of civilizations with that of our species. Humans have been around for at least a million years, but there have been thirty civilizations in the past five thousand years alone. Humans are tough and adaptable; civilizations fragile and short-lived.

We are proud of our "intelligence," but there is little evidence that it has improved through all the years of recorded history. Our intelligence does not seem to be evolving; we do not seem to becoming true *Homo sapiens*. We are more, as E. O. Wilson put it, like unfortunate tribal carnivores who have acquired intelligence. Our evolution is more like that of social insects—the knowledge and understanding we have gained and value so highly is more a property of our "nest"—our civilization—than its individual members. The nest is always more powerful than any individual. We might swat a solitary hornet, but who dares disturb the hornet's nest? Tiny bees can destroy the large and strong, but solitary, Japanese hornet when it threatens their nest. They cluster around it closely and cook it to death with their body heat. And not even its large brain can save the sperm whale when it is surrounded by possibly less-intelligent human hunters in a group.

We are surprisingly ignorant and inept as individuals. How many of us, lost in a wilderness, could manage to make a flint knife or spear? Who of us knows even a small percentage of everything there now is to know in science? How many of those working in the computer industry could make any of its essential equipment, such as a computer screen or keyboard? The social insects carry the instructions for their nest building in their genes. But we have no permanent record of our civilization, internal or external, from which to restore it, should it fail. We would have to start again from scratch.

Creatures that may have to endure desiccation often enclose their genes in spores, so that what they need to renew themselves will survive any drought. Might we also somehow encapsulate the information essential to our civilization to assure that it survives any dark times?

I often enjoy walking in the moorland near my home. It's easy to get lost here when it gets dark and the mists come down. To avoid getting lost, I make sure that at all times know where I am and how I got there. Our journey into the future is somewhat like this. We may not always be able to see the road ahead or all of the pitfalls in it, but it would help to know our present position and how we got here. For this we need a record that contains all of the key information, written clearly and simply so that anyone can understand it, and carefully kept up to date.

We have no such record now. Most of what the average person knows about the earth, our civilization, or anything else comes from books, magazine articles, and television programs that present either the single-minded views of specialists or the seductive arguments of lobbyists. We live in adversarial times, and tend to hear mainly the views of special-interest groups. None of these are likely to admit that they might be mistaken about anything, and they all fight for the benefit of their own group rather than humankind. What use would an "instruction book" written like this be to the survivors of a future flood or famine? When they dig it from the debris, they would want to know what went wrong and why. What help would they get from the press release of an international oil company, or of a Right to Life lobbyist? Or for that matter, from a governmental report, when science itself has to lobby for support? Those future survivors, too, would find the language of modern science almost incomprehensible. Scientific papers are so packed with jargon that scientists themselves can often understand only members of their own specialty.

Take a good look around in any bookstore or library. The books there may be well written, entertaining, or informative, but most of them deal with superficial and rather evanescent topics. They take a great deal for granted—the hard-won scientific knowledge that gave us the safe and comfortable lives we enjoy today. Most of us are so ignorant of the facts upon which science and our scientific culture are founded that we give equal place on our bookshelves to nonsense like astrology, creationism, and junk science. At first, such things served just to entertain us, or indulge our curiosity—we didn't take them seriously. Now they are often accepted as fact, and given "equal time." Imagine the survivor of an Earth catastrophe trying to cope with a cholera epidemic, with only a book on aromatherapy for guidance. Such a book, alas, would be more likely to be found amid the debris of our present society than an easily understandable medical text!

A permanent record of the underpinnings of our civilization may not be as hard to put together as it sounds. What we need is a primer on science, clearly written and unmistakable in its meaning—a guidebook for anyone interested in the state of the Earth and how to survive and live well on it. It would be the scientific equivalent of the Bible. It would explain the evolution, or natural selection of living things, and what we know about the universe. It would explain the basic principles of medicine and surgery, how the blood circulates, and the functions of the different organs. We take the existence of bacteria and viruses, and the facts of pharmacology, all for granted, too, but this precious knowledge could easily be lost and take centuries to recover. The same is true of the facts of engineering and thermodynamics. The history of chemistry reveals how long it took to discover the periodic table of the elements, and to unravel the mysteries of the air, the rocks, and the oceans. Here things like this would all be clearly recorded and preserved. This book would include practical information as well, such as how to light and maintain a fire and how to tap an aquifer.

The book I am proposing would be more than a survival manual. It could also serve as an elementary school text, presenting science to schoolchildren in a relevant and interesting way.

This vital volume would not be one for the magnetic or optical media, or for that matter any medium that needs a computer or electricity to read it. Words stored in such a form cannot be counted upon. These media themselves are short-lived and vulnerable, and nothing stored in them can be read without specialized hardware and software. They require sophisticated technology we cannot assume will be always available. And rapid obsolescence is the norm here. Modern media are less reliable instruments for long-term storage than the original culture record, the spoken word.

What we need is a book, written on durable, long lasting paper, and sturdily bound. The information in it must be clear, unbiased, accurate, and up to date. Above all, we all need to accept it and believe in it at least as much as we in the United Kingdom believed in, and perhaps still do believe in, the World Service of the BBC.

After the disintegration of the Roman Empire, monks were the protectors of knowledge that society had lost interest in or was hostile to. Much of this knowledge was in books, which the monks preserved and read. Modern scientists, ever more specialized, do not feel this obligation, and science—with the exception of a few isolated institutions like the National Center for Atmospheric

Research—has no counterpart to monasteries. There are no guide-lines now—the book would have to stand on its own. If well written by the most authoritative and respected individuals, it might need no protectors. It would, like the Bible, find its place in every library, school, home, and house of worship. It would then be available and ready for when it would be needed most.

Contributors

James E. Alcock is Professor of Psychology at York University in Toronto, Ontario, Canada. His research interests lie in the psychology of belief systems, with a particular focus on formation of belief in extraordinary phenomena. He is the author of two critical books dealing with psychology and the paranormal, *Parapsychology: Science or Magic?* (1981) and *Science and Supernature* (1990), co-editor of *Psi Wars* (2003), co-author of a textbook of social psychology, and author of nine book chapters and numerous articles and papers. He is a fellow of the Canadian Psychological Association and member of: American Psychological Society; American Psychological Association; College of Psychologists of Ontario; editorial board, *Skeptical Inquirer;* advisory board, American Council on Science and Health; Council for Scientific Medicine; editorial board of *The Scientific Review of Alternative Medicine;* Council for Scientific Clinical Psychology and Psychiatry.

Paul Baer is a graduate student at the University of California, Berkeley working with Richard Norgaard.

Andy Bahn is a graduate student at Rensselaer Polytechnic Institute in Troy, New York working with John Gowdy.

Jerome H. Barkow is a sociocultural anthropologist, teaching and researching in evolution and human nature and anthropologies of food and of health at Dalhousie University in Halifax, Nova Scotia. He is currently conducting research in west Africa, Nova Scotia, and Indonesia, comparing production of indigenous knowledge with production of scientific knowledge. A list of his selected publications includes: "Introduction" and "Beneath new culture is old psychology" in *The Adapted Mind: Evolutionary Psychology and the Generation of Culture* (J. Barkow, L. Cosmides, & J. Tooby, eds., 1992); "Social competition, social intelligence, and why the Bugis know more about cooking than about nutrition" (lead author) in *The Origins of Human Social Institutions* (W.G. Runciman, ed., 2001, in the *Proceedings of British Academy series*); *Missing the Revolution: Evolutionary Psychology for Social Scientists* (J. Barkow, ed., forthcoming).

John M. Gowdy has degrees in anthropology and community planning, and a Ph.D. in economics, from West Virginia University. He is Rittenhouse Professor of Humanities and Social Science, Department of Economics, professor of economics and director of the PhD program in ecological economics at Rensselaer Polytechnic Institute in Troy, New York. His research interests include: ecological economics, economic anthropology, input-output analysis of energy and resource use, and evolutionary models of economic change. Dr. Gowdy has been a visiting scholar at universities in Europe, Japan, and Australia, and holds membership on the editorial boards of various journals including: *Ecological Economics, Structural Change and Economic Dynamics,* and *Environmental Ethics.* He has edited or authored nine books and published over 80 academic articles. His most recent book is *Paradise for Sale: A Parable of Nature,* co-authored with Carl McDaniel and published by the University of California Press.

Lindsey Grant writes on population and public policy. A retired foreign service officer, he has served as Director of the Office of Asian Communist Affairs, a National Security Council staff member, and Department of State Policy Planning Staff member. As Deputy Assistant Secretary of State for Environment and Population Affairs, he was Department of State coordinator for the Global 2000 Report to the President, Chairman of the Interagency Committee on International Environmental Affairs, U.S. delegate to (and Vice Chairman of) the OECD Environment Committee, and U.S. member of the UNECE Committee of Experts on the Environment. His books include: *Foresight and National Decisions: the Horseman and the Bureaucrat* (1988); *Elephants in the Volkswagen: A Study of Optimum U.S. Population* (1992); *How Many Americans?* (with Leon Bouvier) (1994), *Juggernaut. Growth on a Small Planet* (1996); *Too Many People. The Case for Reversing Growth* (2001); and *The Collapsing Bubble: Growth and Fossil Energy* (2005).

John H. Herz is Professor Emeritus of Political Science, City College and Graduate School, The City College of New York, and has also taught at Howard University, Columbia University, Trinity College, The New School, Phillips-Universitat Marburg, and the Free University of Berlin. He is a fellow of the Graduate Institute of International Studies in Geneva, Switzerland and a former member of the Institute of Advanced Studies, Princeton, New Jersey, and has contributed such basic concepts to political science as "security dilemma," "territoriality," "impermeability," and "penetrability."After having become aware of new 20th century threats to civilization and our planet, Dr. Herz devoted himself to defining and drawing attention to these threats. Selected publications include: *Political Realism and Political Idealism* (1951); *International Politics in the Atomic Age* (1956); *The Nation-State and the Crisis of World Politics: Essays on International Politics in the Twentieth Century* (1976); *Vom Überleben: wie ein Weltbild entstand: Autobiographie (On Human Survival: How a World-View Emerged: Autobiography)* (1984).

Richard D. Lamm is Co-Director of the Institute for Public Policy Studies. He argues that the challenge of the 21st Century is to meet new public needs by reconceptualizing much of what government does and how it does it. He was selected as one of *Time* magazine's "200 Young Leaders of America" in 1974, and was chairman of the Pew Health Professions Commission and a public member of the Accreditation Council for Graduate Medical Education. Selected publications include: *Megatraumas: America in the Year 2000* (1985); *The Immigration Time Bomb: The Fragmenting of America* (with Gary Imhoff) (1985); and *The Brave New World of Health Care* (2003).

Ervin Laszlo has a PhD from the Sorbonne and is the recipient of four honorary PhD's, the Japan Peace Prize (Goi Prize), and other distinctions. He was nominated for the Nobel Peace Prize in 2004 and was re-nominated in 2005. Formerly a professor of philosophy, systems science, and futures studies in various universities in the U.S., Europe, and the far east, Dr. Laszlo is the author or co-author of forty-five books translated into as many as twenty languages, and the editor of another twenty-nine volumes, including a four-volume encyclopedia. He is founder and president of The Club of Budapest, founder and director of the General Evolution Research Group, president of the Private University for Economics and Ethics, fellow of the World Academy of Arts and Sciences, member of the International Academy of Philosophy of Science, senator of the International Medici Academy, and editor of the international periodical *World Futures: The Journal of General Evolution*.

James Lovelock holds a PhD in medicine from the London School of Hygiene and Tropical Medicine; and a DSc in biophysics from London University. Since 1964 he has conducted an independent practice in science and has filed more than 50 patents. Dr. Lovelock is the originator of the Gaia Hypothesis (the Earth behaves as if it were a super organism, made up from all the living things and from their material environment), and is author of approximately 200 scientific papers, distributed almost equally among topics in medicine, biology, instrument science, and geophysiology. He is also author of four books on the Gaia Hypothesis and an autobiography, *Homage to GAIA: The Life of an Independent Scientist* (2000). Dr. Lovelock was made a C.B.E. (Commander of the British Empire) by Queen Elizabeth II in 1990.

Walter Lowen holds a PhD in nuclear engineering from the Eidgenossische Technische Hochschule, Zurich, Switzerland, and an M.S. in mechanical engineering from North Carolina State University in Raleigh. He is Professor Emeritus of systems science at the Thomas J. Watson School of Engineering and Applied Science, State University of New York-Birmingham, where he joined the faculty as dean of the School of Advanced Technology. Dr. Lowen has been an invited sabbaticant at IBM's Systems Research Institute and a consultant to Alco, Oak Ridge National Laboratory, the Knolls Atomic Laboratory, and the Institute of Applied Technology of the National Bureau of Standards. HE has published numerous articles and papers and, with Lawrence H. Miike, authored the landmark model of cognitive skills, *Dichotomies of the Mind: A Systematic Explanation of Human Behavior* (1982).

J.R. McNeill was educated at Swarthmore College and Duke University, and since 1985 has been teaching history in the Walsh School of Foreign Service at Georgetown University, where he also holds the Cinco Hermanos Chair in Environmental and International Affairs. He has held two Fulbright awards, a MacArthur grant, and a Woodrow Wilson Fellowship. He is the author or editor of six books, including *The Mountains of the Mediterranean World: An Environmental History* (1992); *Something New Under the Sun: An Environmental History of the 20th-Century World* (2000), and *The Human Web: A Bird's-eye View of World History* (with William H. McNeill) (2003).

David G. Myers holds a PhD from the University of Iowa and is the John Dirk Werkman Professor of Psychology at Hope College in Holland, MI. His scientific writings, in such journals as *Science, the American Scientist, the American Psychologist,* and *Psychological Science,* have garnered the support of the National Science Foundation and been recognized with the Gordon Allport Prize. In some 400 media interviews and invited lectures, he has challenged America's individualism and materialism and affirmed the significance of positive traits, committed relationships, and religious faith. His selected publications include: *Who Is Happy—and Why* (1993), *The American Paradox: Spiritual Hunger in an Age of Plenty* (2000), *Psychology* (1986-

2003, with editions in nine languages), *Intuition: Its Powers and Perils* (2004).

Richard B. Norgaard earned his PhD in economics from the University of Chicago, and holds an MS in agricultural economics from Oregon State University. Currently a professor at the University of California, Berkeley, Dr. Norgaard is considered one of the founders of the field of ecological economics. His recent research addresses how environmental problems challenge scientific understanding and the policy process, how ecologists and economists understand systems differently, and how globalization affects environmental governance. The author of one book and co-author or editor of three more, he has over 100 other publications spanning the fields of environment and development, tropical forestry and agriculture, environmental epistemology, energy economics, and ecological economics. Dr. Norgaard is a member of the Environmental Economics Advisory Committee of the Science Advisory Board of the U.S. Environmental Protection Agency, serves on the Board of the American Institute of Biological Sciences, and has served as president of the International Society for Ecological Economics.

David Pimentel holds a PhD from Cornell University in Ithaca, New York and conducted his postdoctoral research at Oxford University, the University of Chicago, and MIT. His is a professor of ecology and agricultural science at Cornell University whose research is focused on the theoretical and experimental aspects of the population dynamics and coevolution in parasite-host systems. This basic research has led to both theoretical and experimental research on the biological control of pests. He has served on many national and government committees, including the National Academy of Sciences, the President's Science Advisory Council, the Office of Technology Assessment of the U.S. Congress, and has worked directly with the U.S. State Department. He has published more than 500 scientific papers and 20 books.

Marcia Pimentel is a senior lecturer in the Division of Nutritional Sciences, College of Human Ecology, Cornell University. She has published 30 scientific papers and the following books: *Food, Energy, and Society* (1996) and *Dimensions of Food* (1999).

Peter Seidel was a student of renowned architect Mies van der Rohe and planner L. Hilberseimer. He earned his MS at the Illinois Institute of Technology and has taught at universities in North America, China, and India. As a reaction to architecture popularizing energy wasting glass office towers, Mr. Seidel became a systems oriented environmental planner and architect. A list of his selected publications includes: *Invisible Walls: Why We Ignore the Damage We Inflict on Our Planet and Ourselves* (1998); numerous articles on environmentally friendly buildings and sustainable community planning in *World Futures, The Futurist,* and *the Indian Journal of Applied Economics;* and the special double issue of *World Futures* on Survival Research—where much of this volume first appeared—of which he was guest editor.

Joseph Tainter is a research professor in the International Institute for Sustainability at Arizona State University, Tempe. He is the author of *The Collapse of Complex Societies* (1988); co-editor of *Evolving Complexity and Environmental Risk in the Prehistoric Southwest* (1996) and *The Way The Wind Blows: Climate, History, and Human Action* (2000); and co-author of *Supply-Side Sustainability* (2003).

Kenneth E. F. Watt is Professor Emeritus of evolution and ecology at the University of California, Davis. He earned his PhD in zoology from the University of Chicago and received an honorary LL.D. from Simon Fraser University (Burnaby, B.C., Canada) for his work on mathematical models of animal systems and human societies. His books are published in Russian, Japanese, and Spanish, and he has given lecture courses to classes of professors in Caracas, Venezuela and Beijing, China. He was one of seven outside advisors to The Global 2000 Report to President James Earl Carter. He believes a central problem for the U.S. at present is its intra-disciplinary, rather than interdisciplinary, approach to knowledge.

Dr. Christopher Williams is currently at the Centre for International Education and Research, University of Birmingham (U.K.). He is a former instructor at the United Nations University Leadership Academy (UNULA) in Amman, Jordan, as well as at the universities of Cambridge, London, and Cairo. While a senior fellow of the Economic & Social Research Council Global Environmental Change Programme, he wrote *Terminus Brain: the Environmental Threat to Human Intelligence,* (republished in German by Klett-Cotta), and *Environmental Victims: New Risks, New Injustice.* As a Joseph Rowntree Fellow, his work included *Invisible Victims: Crime and Abuse Against People with Learning Disabilities.* At the U.N. University he wrote *Leaders of Integrity: Ethics and a Code for Global Leadership.* His latest book is *Leadership Accountability in a Globalizing World.*

Index